IMAGES
of America

FLOWING WELLS

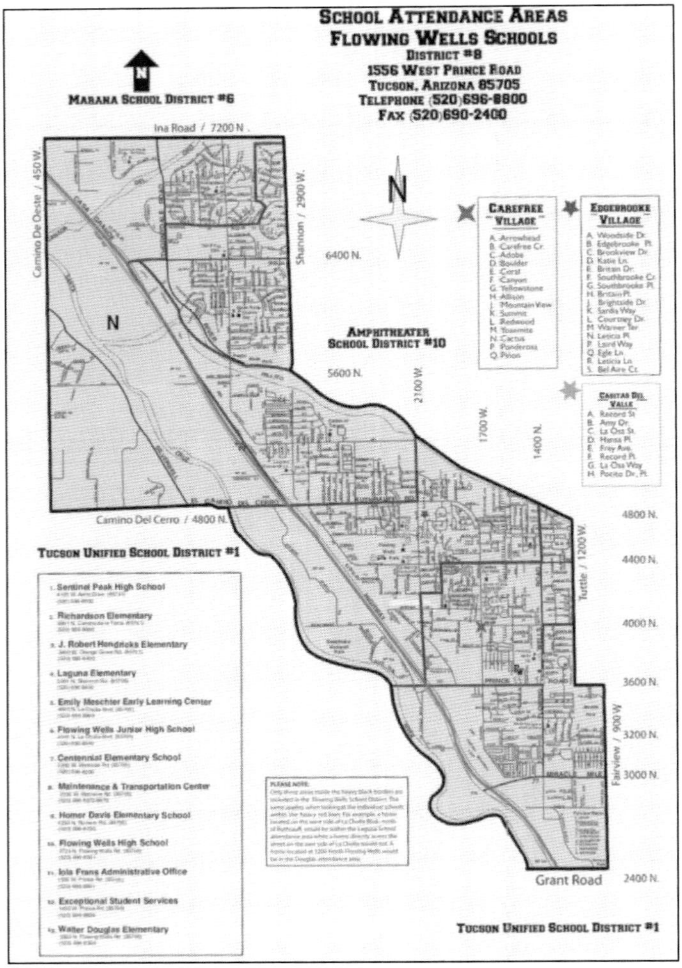

This map shows the modern-day Flowing Wells, approximately 13 square miles and roughly bounded by Grant Road to the south, the Santa Cruz River to the west, Fairview Road to the east, and Ina Road to the north. The area south of Ruthrauff Road to Grant Road is Flowing Wells's historic boundaries. The school district originally named Rillito No. 22 had largely the same boundaries as modern-day Flowing Wells when formed in 1889. In that year, the Rillito School District split into Rillito No. 22 and Laguna No. 36. At the time, the Flowing Wells Ranch and Laguna area had a schoolhouse. In 1899, Rillito changed to district No. 8 and Laguna to No. 17. Rillito's name was changed to Flowing Wells in 1928. In 1964, the Laguna School District was annexed into Flowing Wells, restoring the larger historical district. (Flowing Wells Unified School District.)

ON THE COVER: Laguna schoolchildren do the broom dance, or the waltz of the broom, in the schoolyard with their teacher, Thelma Voris, around 1939. In the dance, there must be an odd person to dance up and down with the broom in center. Music plays for several measures; all stand still except the person dancing with the broom between the lines of boys and girls. The dancer suddenly drops the broom and grabs a partner, which is a signal for all to get a partner. The odd person left over picks up the broom, and the game repeats. The Catalina Mountains are in the background. (Arizona Historical Society.)

IMAGES
of America

FLOWING WELLS

Kevin and Marie Daily

ARCADIA
PUBLISHING

ISBN 978-1-4671-2715-8

Published by Arcadia Publishing
Charleston, South Carolina

Printed in the United States of America

Library of Congress Control Number: 2017938401

For all general information, please contact Arcadia Publishing:
Telephone 843-853-2070
Fax 843-853-0044
E-mail sales@arcadiapublishing.com
For customer service and orders:
Toll-Free 1-888-313-2665

Visit us on the Internet at www.arcadiapublishing.com

This book is dedicated to all Flowing Wells family and friends, past, present, and future, who share a vision of a tight-knit community bigger than any one individual and built on hard work, with the school district at its core. It is a special place where potential finds opportunity. "There's just something special about Flowing Wells."

Contents

ACKNOWLEDGMENTS

The Flowing Wells Neighborhood Association and Community Coalition will donate all royalties earned from sales of this book to fund scholarships for Flowing Wells students.

A special thank-you to Patricia J. Peterson (Kevin's mother) and Maynard "Pete" H. Peterson (Kevin's stepfather) for making it possible for us to learn, through research for this book, the depth of our connection to Flowing Wells through Maynard back to the 1890s via Maime Bouschet and Henry Peterson (Maynard's uncle—Maynard was not aware of the depth of his connection). Kevin graduated from Flowing Wells High School (FWHS) in 1983. He and Marie planned for their children to also experience this special place where potential will find opportunity and become part of the Flowing Wells family.

Many thanks go out to the organizations and individuals that were vital in supplying images and information for this book. These include the Special Collections Library at the University of Arizona, the Arizona Historical Society, Arizona Memory Project, US Department of Agriculture Natural Resource Conservation Service (USDA NRCS), Pima County Maps, the *Arizona Daily Star*, *Tucson Citizen*, Flowing Wells Irrigation District, James and Janet Elliott, Diane Graham, Patricia J. Peterson, Maynard "Pete" H. Peterson, Flowing Wells Unified School District (FWUSD), Eddie Lowry, Sue Rhude, Jeretta Douglas, Jean and John O'Hanlon, Loraine Gerovec, Venla Mitchell, Geraldine "GeeGee" Gee Larrington, John Lawson, Burl Young, Marcia and Roy Claridge, Roger Gardner, Jack Mead, the W.T. Dudgeon family, Kenneth "Ken" Petrick, Douglas Randolph "Randy" Holladay, William Woods Jr., Lynette Christian, Melanie Black, Rudy and Rachel Reuser, the Abrams family, Gary Waters, R.E. Darling Company, David Baker ("potential finds opportunity"), and the Flowing Wells Neighborhood Association and Community Coalition Board. Additional thanks to James "Jimmy" Elliott (FWHS class of 1958) for all his hard work and contributions researching maps, hand-drawing maps of the canal routes, proofreading, and providing an edited version of the introduction that hits a home run. Thanks to Rudy Reuser for his amazing memory and stories.

Another heartfelt recognition goes out to Mary Holloway for her undying love and dedication to the Flowing Wells School District and for being a founder and the go-to information source for alumni of FWHS for decades. To the many residents of the Flowing Wells area: thank you for preserving history in the images passed down through the generations. To our editor Caitrin Cunningham and Arcadia Publishing, thank you for your unwavering support and guidance in helping us share the history of Flowing Wells.

INTRODUCTION

The Flowing Wells School District covers 13 square miles, partially in the city limits of Tucson as well as in unincorporated Pima County, Arizona. The area, the water district, and the school district attribute their name to the historic Flowing Wells Ranch and its source of water originally established by brothers Frank and Warren Allison in 1896.

It is helpful to remember that the primary sources of water in our area in times gone by were wind-driven well pumps and diversion ditches from up-river banks. Electricity generation hadn't been invented yet, let alone implemented in places as remote as Tucson in the Territory of Arizona. In those times, the Santa Cruz River flowed from south to north passing between early Tucson and the base of a hill called Sentinel Peak (now A Mountain). In the late 1800s, the river in this area had been described as a clear, slow moving, wide and shallow, year-round stream that overflowed its banks nearly every monsoon season. Water for domestic, farming, and recreational use was collected, stored, and transported from this location to the townspeople and surrounding farmers.

As the population and farm production grew, the river flow was becoming less and less adequate. Two enterprising brothers named Frank and Warren Allison, while attempting to drain and recover some swampland near Sentinel Peak, discovered almost by accident the existence of a subterranean river beneath the Santa Cruz Riverbed referred to as the "underflow." They tapped into this underflow by driving a series of 10-inch pipes to a depth of about 10 feet below the riverbed and produced a very large increase in usable water flow. These man-made artesian wells were called the Flowing Wells.

Following that good fortune, the Allison brothers acquired rights to a considerable amount of the Santa Cruz River's surface and underflow waters on July 22, 1895. Around 1896, they acquired about 500 acres of uncultivated government land approximately three miles north of downtown Tucson and began construction of a large canal to bring the water to that land. Those 500 acres are believed to have been roughly bounded by today's Prince Road on the north, Miracle Mile on the south, Fairview on the east, and Romero Road on the west. The canal was completed in 1897 and was initially known as Allison's Ditch. The brothers called their land the Flowing Wells Ranch, and it included a water wheel–powered grinding mill near today's Barrio Anita area of Tucson.

The entire operation (wells, canal, and ranch) was sold to Levi Manning and associates for $60,000 in 1902. They further developed the canal, ranch, and water supply headworks, which also supplied the growing City of Tucson. Manning was responsible for installing a series of deep interconnected wells across the Santa Cruz that supplied Tucson. This became known as the "Cross Cut." The Allison canal became known as "Manning's Ditch."

In 1911, the entire operation was sold again, this time to a group of investors from out of town who formed the Tucson Farms Company. The company divided the land into what it advertised in the Midwest and elsewhere as ready-to-farm plots with abundant water. In 1921, farmers in the area who were not happy with the quantity and pricing of water petitioned the Pima County Board of Supervisors to create their own water district, which subsequently led to the creation of two water districts: the Flowing Wells Irrigation District (FWID) on the south side of the Rillito River, and the Laguna Irrigation District north of the river. Just like Manning and the Tucson Farms Company, the FWID supplied water to Tucson via the Cross Cut wells, some of which now had electric pumps.

In 1940, a disastrous flood damaged the headworks of the canal (by this time known as the Flowing Wells Canal). The FWID deemed it too costly to repair, and abandoned it and the canal completely in favor of sinking only local deep wells with electric pumps. Their interests at the base of Sentinel Peak were sold to the City of Tucson for $16,000 and the canal was gradually filled in, except for those parts that appear to still be a part of northern Tucson's storm drain system.

The Flowing Wells Irrigation District survived the Depression years and the 1940 disaster due to the hard work and dedication of many community members. It thrives today under excellent management and continues as a proud asset of the community. Deep wells and large storage tanks dispersed throughout the local Flowing Wells area adopted after the canal was closed are still the basis of today's water supply system. Little has changed in this regard since then, except that now virtually all the water goes to residential and business use.

The Flowing Wells schools have been a proud asset of the community and a reason why generations of families have stayed in the area. As time passed, the Flowing Wells area changed from large farms and dairies to smaller and smaller farms. During the 1940s after World War II and on into the 1950s and 1960s, many of these were converted to residential and small business uses. After the mid-1960s, narrow dirt roads once lined with large cottonwood trees and irrigation ditches gave way to wider and wider paved streets. Orchards, cotton fields, alfalfa fields, dairy property, and even an airport became business and residential properties. Throughout its history, the school district has been the centerpiece binding the Flowing Wells community. People who work with other schools and communities have been quoted as saying, "there is just something special about Flowing Wells! We wish it could be bottled up and duplicated elsewhere." This is the story of Flowing Wells, where a sense of community identity can be traced to its beginning in the 1890s.

One

Water from Sentinel Peak, the Allisons, and Farming

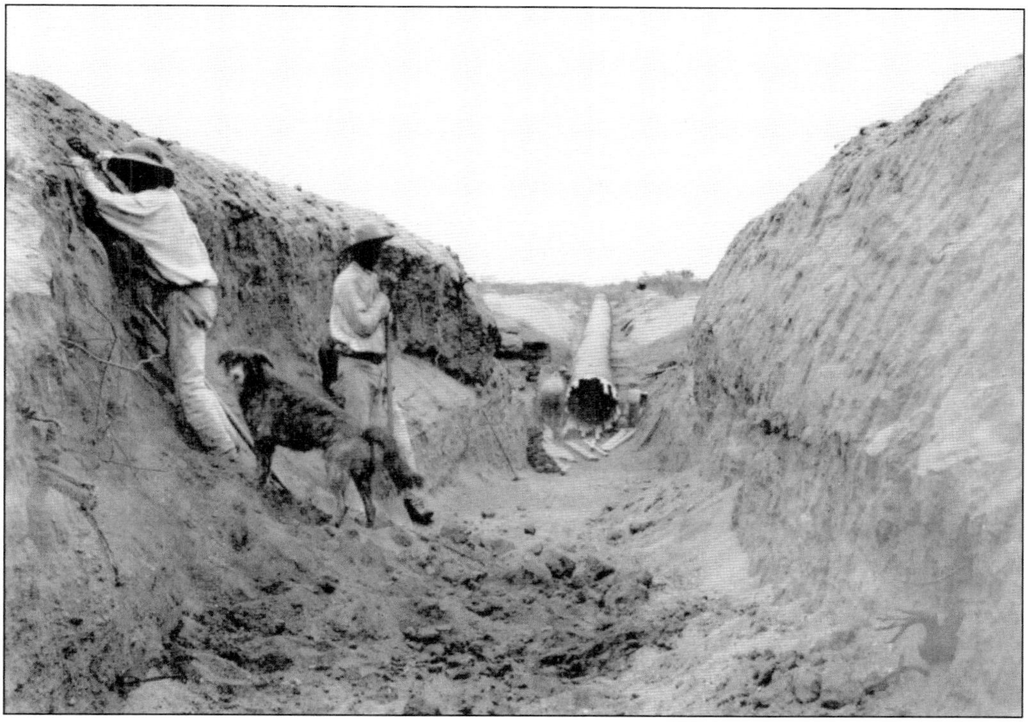

The Rillito Irrigation Project was laid in the 1890s by Alex J. Davidson. Frank and Warren Allison may have used wooden pipes like this to source water to Flowing Wells from Sentinel Peak. Before they dug their ditch, father George and sons Frank and Warren Allison made Tucson their home in 1881. They operated a grocery and lived above the store, but they turned their attention more and more to farming. In the early 1890s, Frank and Warren Allison established a 500-acre farm north of Tucson and acquired the water rights to Warner's Lake at the east base of Sentinel Peak (now A Mountain). George, Frank, and Warren had purchased Warner's Lake in 1886 from Solomon Warner and lived in adobe buildings close to the old Warner's Mill. They stocked the lake with fish and used it for irrigation. They plunged pipes to drain a swamp they were hoping to reclaim, but the three-inch drain pipes to the depth of 10 to 15 feet resulted in several flowing wells. The underflow they discovered was under layers of a hard substance they had to break through. They named their 500-acre ranch Flowing Wells and started building the Allison Canal to channel this new source of water to the ranch. (Arizona Historical Society.)

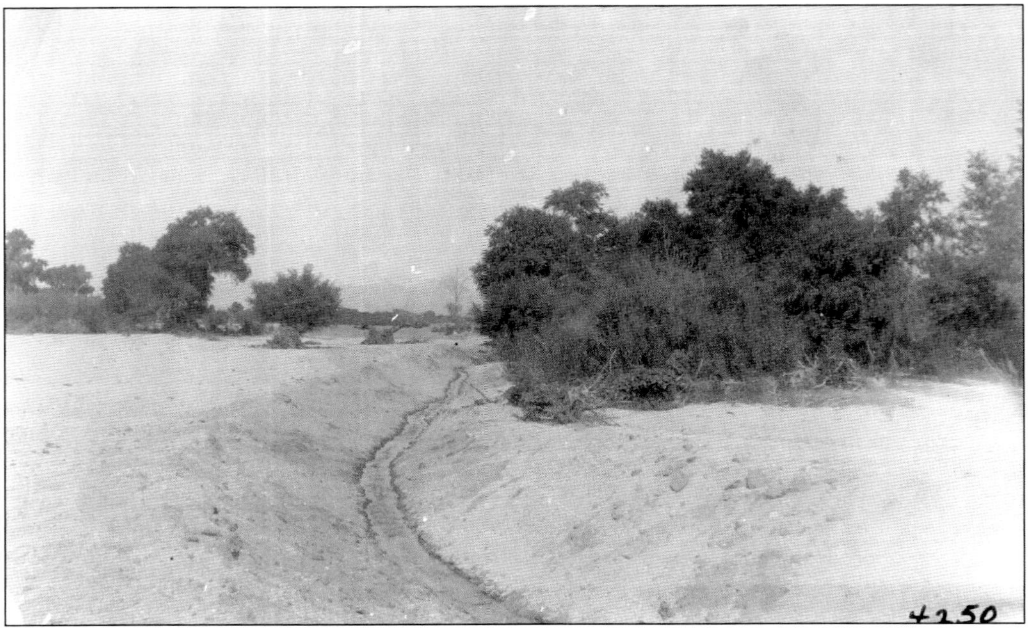

This is an earlier canal built in the 1890s by the Allison brothers on the west side of the Santa Cruz River. It is similar to their Flowing Wells Canal built on the east side of the river in 1896 and 1897. Note the size of the ditch. The Flowing Wells Canal (Allison's Ditch) carried water at a rate of 75 cubic feet per second. (Arizona Historical Society.)

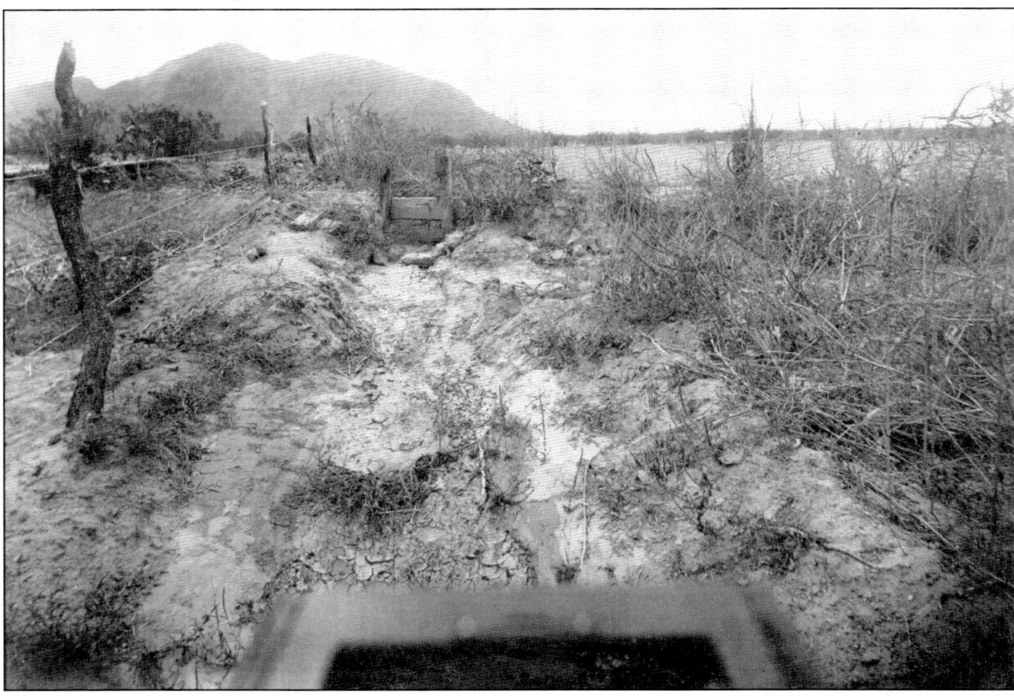

This photograph provides an example of the wooden irrigation gates built to regulate the water flow to farm crops from the irrigation ditch. Tree branches and barbed wire were used to construct the fence. The photograph was taken near what is now Interstate 10 and Avra Valley Road in the early 1900s at the northernmost point of the Tucson Mountains. (Arizona Historical Society.)

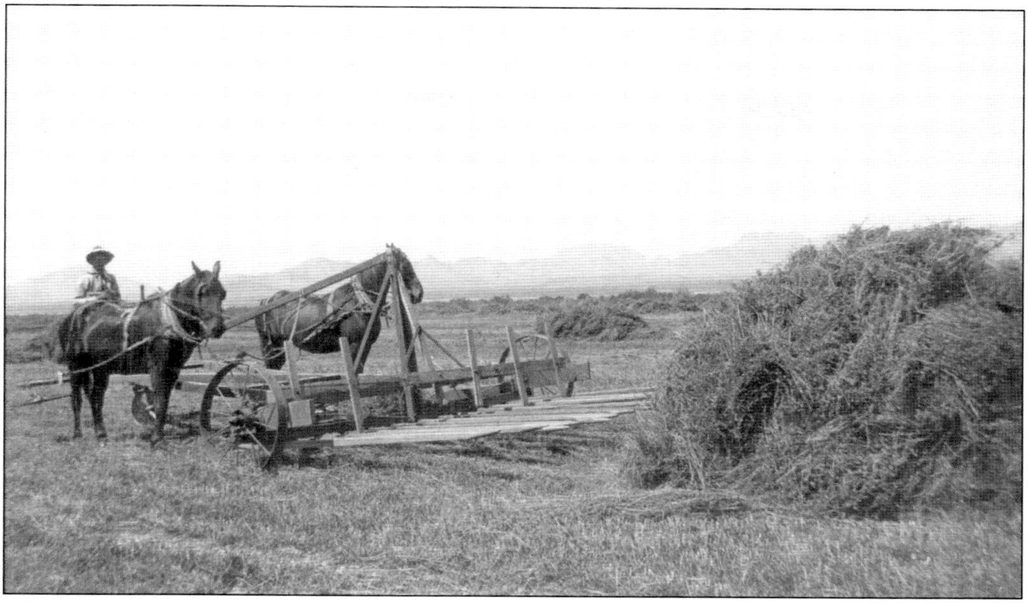

This photograph was taken around 1897 from the Flowing Wells Ranch facing west. The Tucson Mountains are in the background. Alfalfa is being baled the old-fashioned way; it was a major crop, with several cuttings each year. Other crops included watermelons and cantaloupes. The ranch's large watermelon crop was famous, and the melons were shipped by carload to all of southeastern Arizona. The location of this photograph is south of Prince Road and Flowing Wells Road. (Arizona Historical Society.)

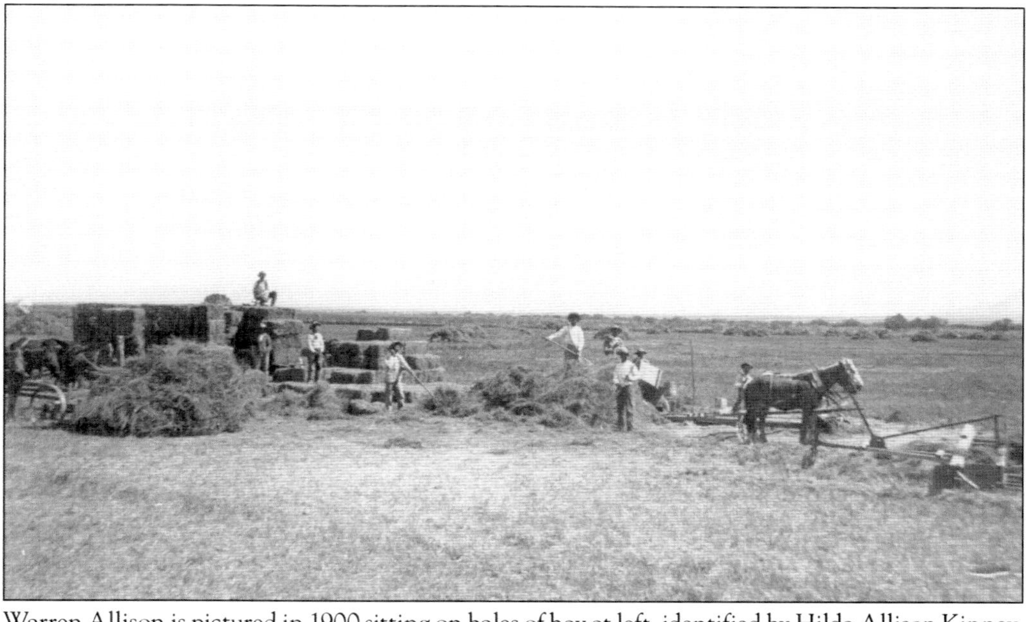

Warren Allison is pictured in 1900 sitting on bales of hay at left, identified by Hilda Allison Kinney, his daughter. The Flowing Wells Ranch produced seven cuttings of alfalfa from April to October. When dried and baled, it sold for $30 per ton, according to Warren Allison's granddaughter Tallia Cahoon. This photograph was taken facing north, with the Tortolita Mountains in the background and the western edge of the Catalina Mountains in view. (Arizona Historical Society.)

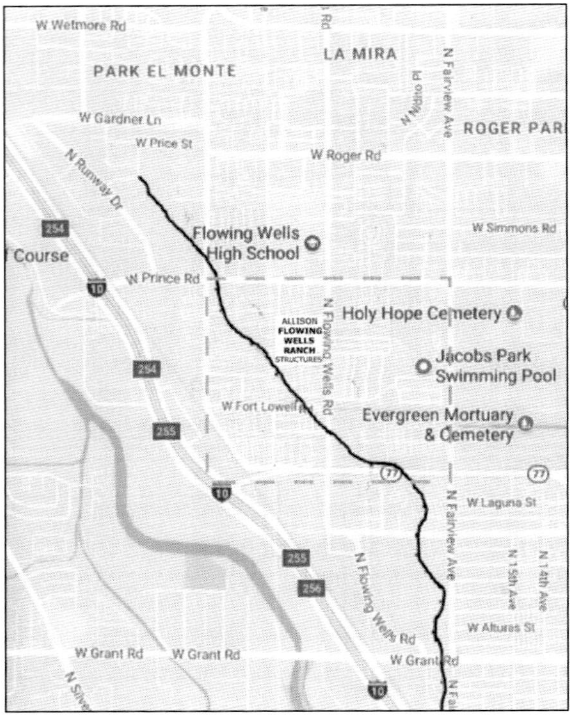

This map shows the estimated path of the original Flowing Wells and Allison Canal of 1896. The section shown is from Grant Road (formerly DeMoss-Petrie Road) into the Flowing Wells Ranch. The ranch is estimated to be on the west side of the 3000 block of modern-day Flowing Wells Road. The canal entered a lake near what is now 1067 West Miracle Mile at the Desert Shores Trailer Park. The lake and a stone lighthouse were part of an upscale 300-seat supper club from the 1930s to 1942. In 1943, it became the 16-acre trailer park under new owners Raymond and Bernice Cox. Their daughter, Lois Skvarek, who grew up at the park, remembers, "I've seen old photographs taken of the place in 1902. That's when it was part of a ranch. The lake was there then; so was the stone lighthouse." For years, KVOA radio broadcast barn dances from the park. (Map by James Elliott, *Arizona Daily Star.*)

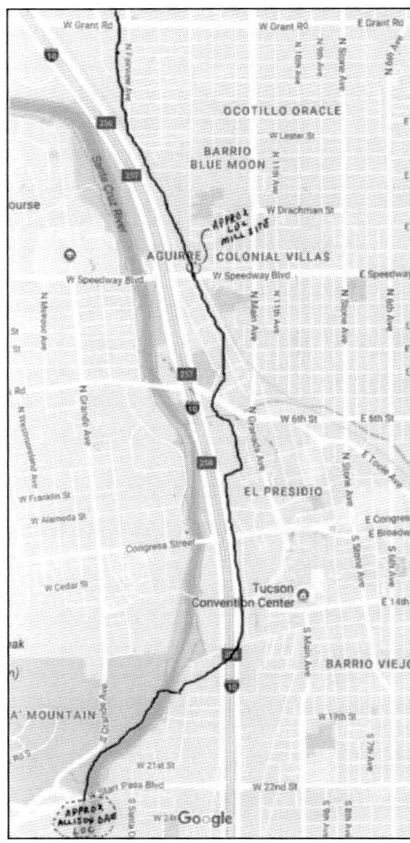

This map was drawn from records of the original canal route from the base of Sentinel Peak to today's Grant Road. Construction of the Allison Dam began in 1893 on the Santa Cruz River near Warner's Lake. The dam was 10 feet high from bank to bank, constructed using heavy timber and a pile driver sent from St. Louis. The lake was four to eight feet deep and three quarters of a mile long. The biggest flood recorded at the time hit in the fall of 1900, destroying Warner's Lake. The Allisons installed the grain mill shown on the map around 1897 in what is now Barrio Anita. (Map by James Elliott.)

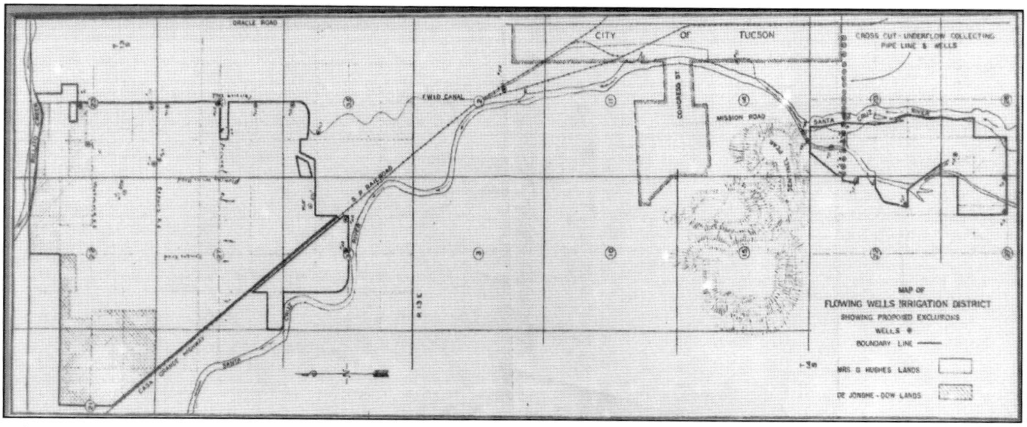

This c. 1922 map depicts the Flowing Wells Irrigation District canal (formerly the Allison Canal and Manning's Ditch) from the base of Sentinel Peak. The crosscut underflow collecting pipeline and wells are at upper right. The canal enters Flowing Wells at Fairview Avenue and Miracle Mile at the lake in the modern-day Desert Shores Trailer Park. Grain and hog farmer Henri F. DeJonghe was a major land holder with about 240 acres on the north end of Flowing Wells, shown at left. After his wife's death around 1918, DeJonghe returned to Chicago and leased this land to Allen Potter, formerly deputy state cattle inspector, and John Schumaker, a Tucson patrolman. (FWID.)

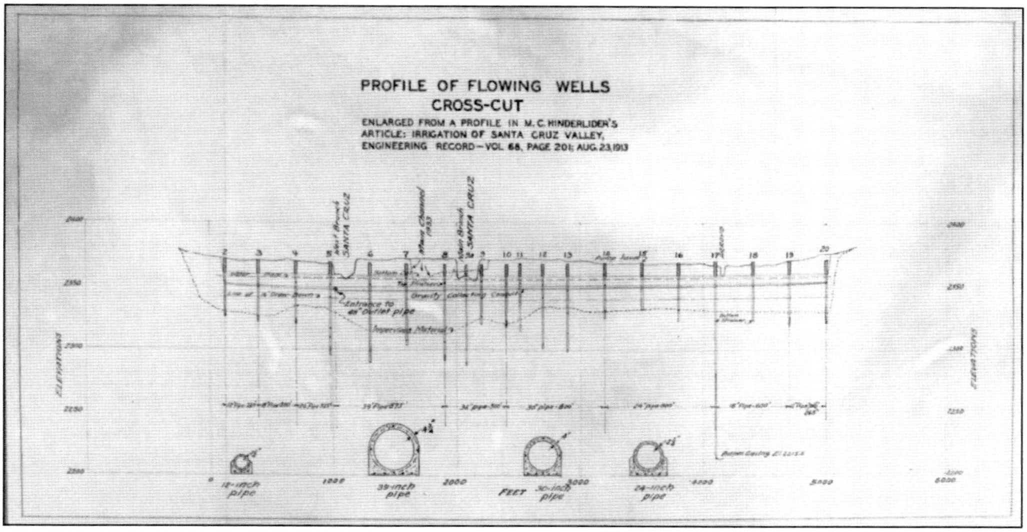

This cross section of the crosscut at the base of Sentinel Peak shows the size, depth, and spacing of the pipes sunk into the ground. Also shown is the hard barrier layer. (FWID.)

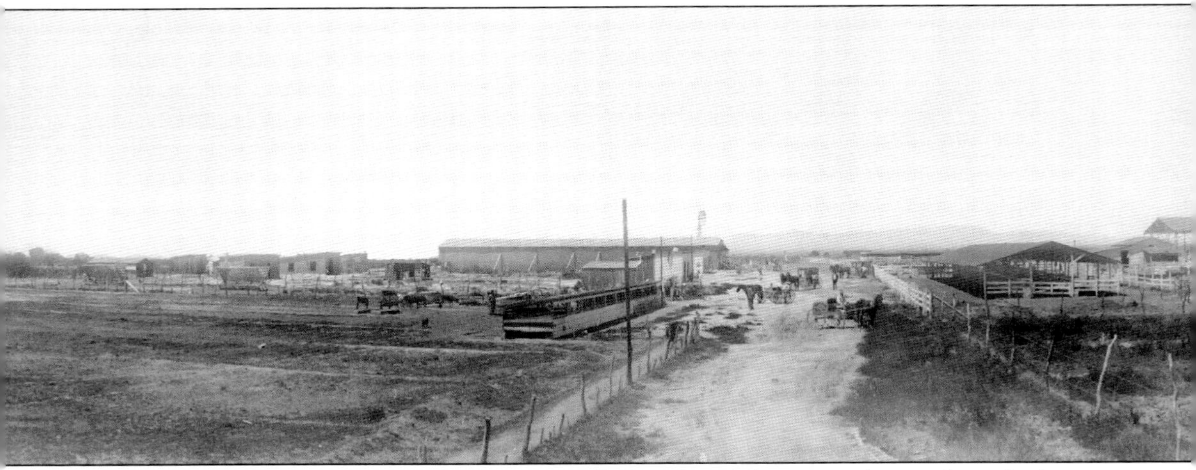

This panoramic photograph of the Flowing Wells Ranch and dairy was taken at about the location of 3300 North Flowing Wells Road facing north. The Tortolita and Catalina Mountains are visible. The words "Flowing Wells Dairy" can be seen on the milk wagon in the center of the road in front. The photograph is believed to have been taken by A.J. Wells and E.L. Huntington, traveling for the bureau of economics of the Southern Pacific Railroad, for a magazine planned for August or November 1912. This picture was mentioned by the Tucson Farms Company in ads to potential

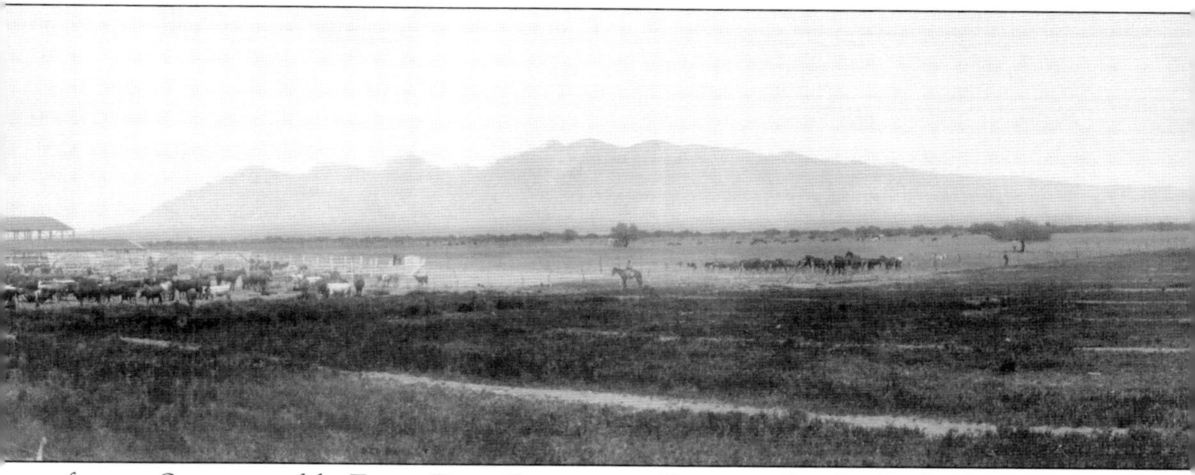

farmers. One owner of the Tucson Farms Company was Chicagoan Douglas Smith. The Flowing Wells area was promoted in 1917 by George Wharton in *Arizona, the Wonderland*: "The Tucson Farms Company owns and operates the largest and most modern dairy in Tucson, known as the Flowing Wells Dairy. The chief object in operating the dairy is to supply to the people of Tucson the best milk obtainable, and incidentally to furnish an outlet for the farmers who have bought lands of the Tucson Farms Company." (Arizona Historical Society.)

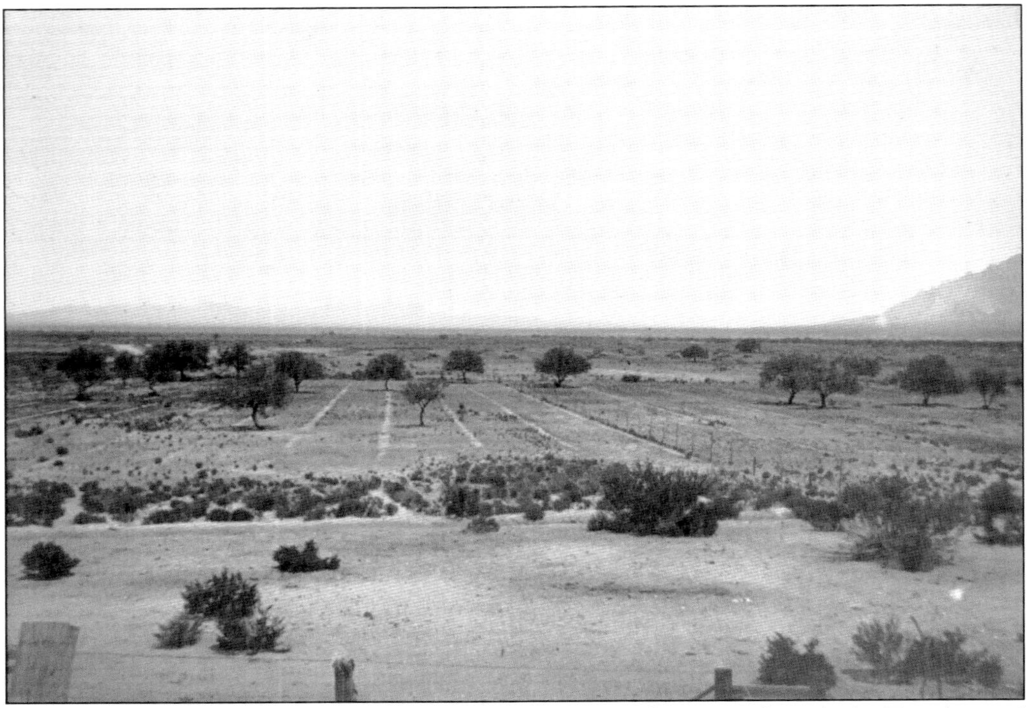

Fields at the Flowing Wells Ranch are pictured around 1900 facing north, with the Tortolita and Catalina Mountains in the background. The Allison brothers sold the Flowing Wells Ranch for $60,000 in 1902 to the newly formed Flowing Wells Company, headed by L.H. Manning, Eugene S. Ives, Brewster E. Randolph (for whom Randolph Park was named), and Collin Cameron. Manning and Ives were very interested in further developing the underflow's potential and believed they could secure aid from the national irrigation fund. (Arizona Historical Society.)

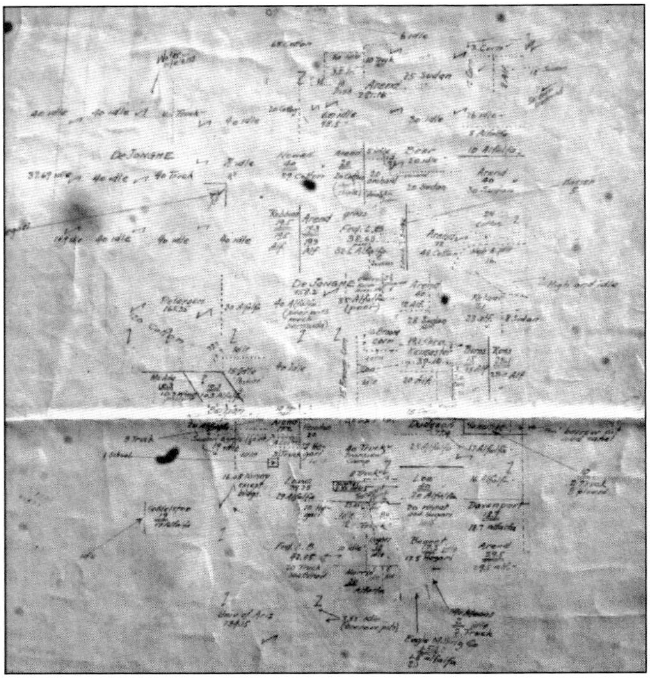

This c. 1933 map shows the land owners and crops planted in Flowing Wells. Some notable owners include Peterson, DeJonghe, Arend, Dudgeon, Hansen, and Pelaar. The transient camp at 3300 North Flowing Wells Road was created around 1933 to house about 225 homeless men in coordination with the Works Progress Administration (WPA) on the original site of the Flowing Wells Ranch. The camp was closed in 1935. David R. Furrey (FWHS 1967) was a long-serving superintendent of the FWID following his father, E.M. Furrey. Furreys Lane and Amy Drive (Amy was E.M. Furrey's wife) are named after the Furrey family. (FWID.)

This c. 1900 photograph of the Flowing Wells Ranch possibly shows bunkhouses, an outhouse, and stable. It was taken facing west with the Tucson Mountains in the background and an alfalfa field in the foreground. (Arizona Historical Society.)

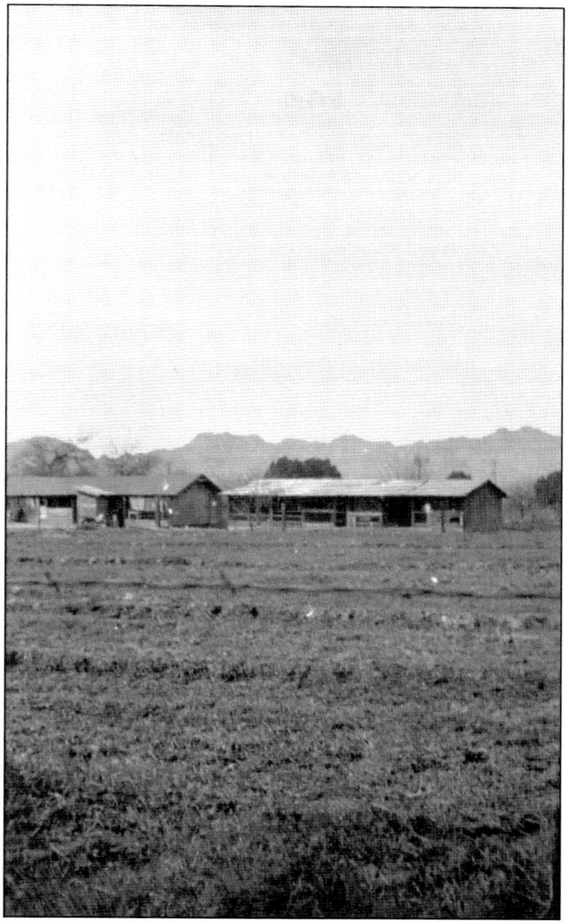

The Flowing Wells Ranch is pictured around 1900 facing southwest with Sentinel Peak in the background, along with a few Tucson buildings. (Arizona Historical Society.)

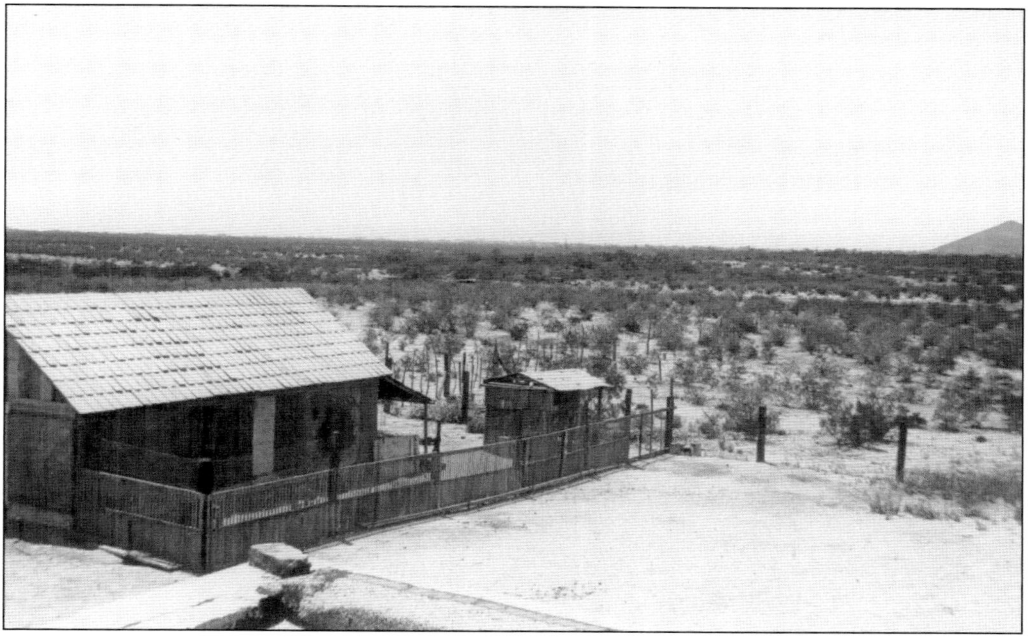

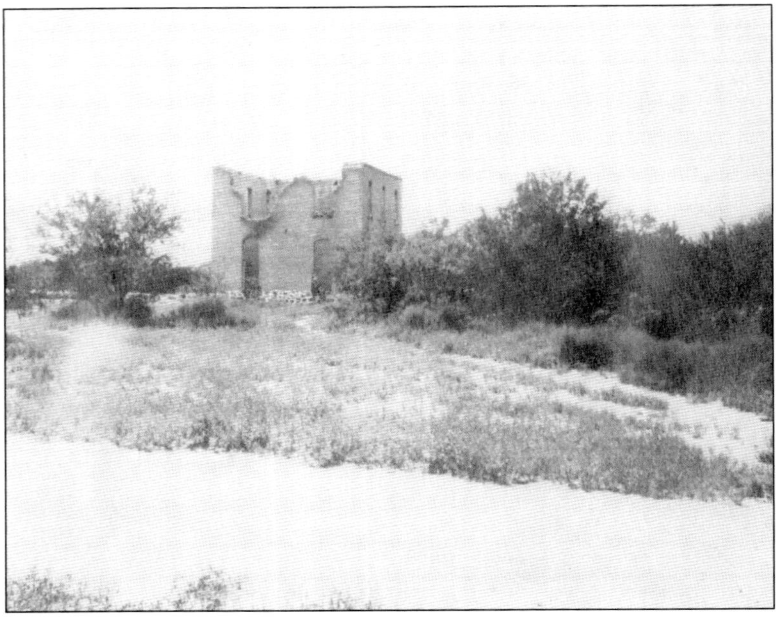

The ruins of the Allison brothers' grain mill are pictured around 1907 in a photograph provided by Hilda Allison Kinney, daughter of Warren Allison. The grain mill was built in 1897 near the present-day Main Avenue and Speedway Boulevard in Barrio Anita. (Arizona Historical Society.)

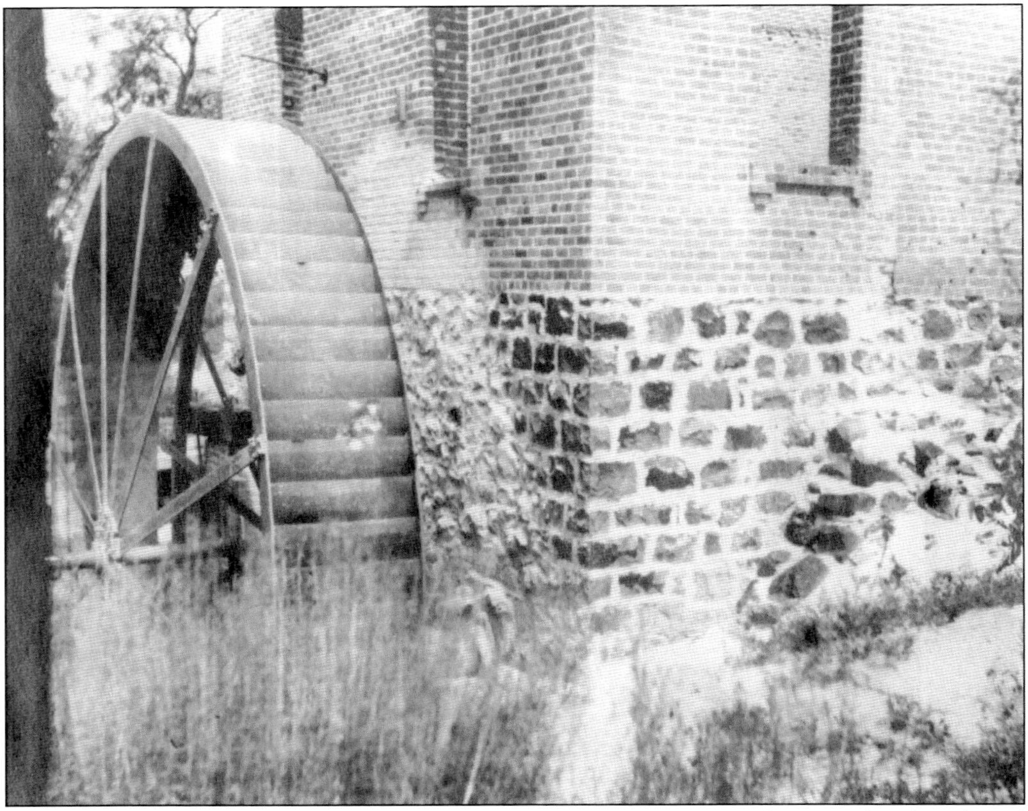

This c. 1907 photograph shows the grain mill with a waterwheel. The mill operated with 60-barrel water power. (Arizona Historical Society.)

As published on December 28, 1911, L.H. Manning sold the Flowing Wells Ranch to the Tucson Farms Company of Chicago, which also purchased another Allison ranch south of Sentinel Peak. The estimated price for both properties was $350,000. Tucson Farms Company planned to search further for the underflow of the Santa Cruz River and to divide the Flowing Wells Ranch into 40-acre tracts to sell to outside settlers for farming. (*Arizona Daily Star.*)

Sale of Allison Ranch Another Step in Proposed Intensive Cultivation of the District, Seeking Underground Flow

The sale of the Allison ranch, of which mention was made in the Star some weeks ago, was but a starter of other land deals of considerable magnitude. The Flowing Wells ranch, owned by L. H. Manning, was purchased yesterday by the same people, Chicago capitalists, who bought the Allison ranch. In addition to these two ranches, options have been procured on most of the land between these ranches. This is but the announcement of what is reported to be a big agricultural proposition in the Santa Cruz valley that forecasts the intense cultivation of the entire Santa Cruz district. Engineers will start to work to determine the water flow and will be able to report definitely in a short time.

The sale of the Allison ranch was a big deal and the purchase of the great Flowing Wells ranch yesterday caused much talk and conjecture. The sale has been pending for some time, the deal being finally closed yesterday. Options on the interlying land have not all been settled as yet but will be within a short time. The consideration has been quoted at $350,000, but Mr. Magenheimer, who is conducting the business for the Chicago syndicate, refused to say what the price was, but admitted it was several hundred thousand.

The Allison ranch is at the head waters of the Allison irrigation ditch, and it is here that the greatest time will be spent by the engineers for the present. The flow of water is known to be fair, but it is thought there is a submerged flow and this will be

Tucson Arizona farms now under a completed

WATER SYSTEM

AND HIGH STATE OF CULTIVATION

Where the Water, Soil, Climate and market conditions are unequalled anywhere in the U. S.

Complied with the Blue Sky Law

The Banking Department of the State of Kansas sent a representative from their office to personally inspect the Tucson Farms Company lands at Tucson, Arizona, and after examining their lands, water system and financial condition of the company issued a permit to S. B. Rohrer to sell their land in the State of Kansas.

8,000 Acres in a High State of Cultivation!

8,000 of our 12,000 acres are in a high state of cultivation. YOU TAKE NO RISK—escape all the chances of the men who pioneered in your locality. On Dec. 15, 1914, wheat, oats and barley were eight inches high—will be harvested in May—and will be followed by a corn crop! We grow more of either crop to the acre than is grown in Illinois or Iowa—and we get better prices. Excellent pasturage all winter—and a forage crop to follow, if you like; all in 12 months.

Crops are a Certainty

From 8 to 10 tons of alfalfa hay yearly to each acre—with prices ranging from $12.50 to $17.50 a ton, baled. And besides the crops we have named, there is big profit in chicken raising (eggs 70c a dozen Dec. 15, never less than 30c). This company is now profitably operating a dairy with 300 cows, but this dairy only partially supplies the local demand. Hog raising, truck gardening and fruit culture pay big, every kind that California can grow, excepting the citrus varieties. Fig, olive and English walnut growing present other great possibilities. Tucson, with 20,000 inhabitants and 10,000 tourists annually, and many nearby cities and towns and numerous cattle ranches and mining camps, are only partly supplied with home grown products. They have until recently imported all of their food supplies—need vastly more produce than these Santa Cruz Valley lands can yield—assure ready cash market for everything you raise.

Splendid neighbors---excellent primary, grammar and high schools, and highly efficient State University and the State Agricultural College.

Let Us Prove to You

That one of these 40 acre farms will pay you from $50.00 to $100.00 net yearly for each acre. The terms so easy that the farm will pay for itself.

Mennonite Settlement

3,600 acres of this land have been reserved for a MENNONITE SETTLEMENT as well as the town site located in the center of the land.

The owners of this project have spent over 2 million dollars to make it a completed water system before offering any of the land for sale.

The Tucson Farms Company is owned and managed by some of the most successful business men in the United States and their reference is any bank in Chicago or Tucson, Arizona.

Free Stereoptican Views of Tucson Farms

Will be given on request of any community. Full information and literature may be had from

TUCSON FARMS COMPANY

S. B. ROHRER,

State Agent and Manager of Mennonite Settlement.

Room 41 Randall Bldg. Phone 1263

NEWTON, KANSAS.

This advertisement for 40-acre tracts in Flowing Wells ran in the *Evening Kansan Republican* on February 13, 1915. Around 1917, settlers complained that Tucson Farms Company was not producing the amount of water promised. This led to the creation of the Flowing Wells Irrigation District. Around 1922, electricity came to rural Pima County, adding pumped water from local wells, but water continued to be delivered in the Allison Canal until 1945. The City of Tucson bought the ditch property and started filling in large sections in 1947. (*Evening Kansan Republican.*)

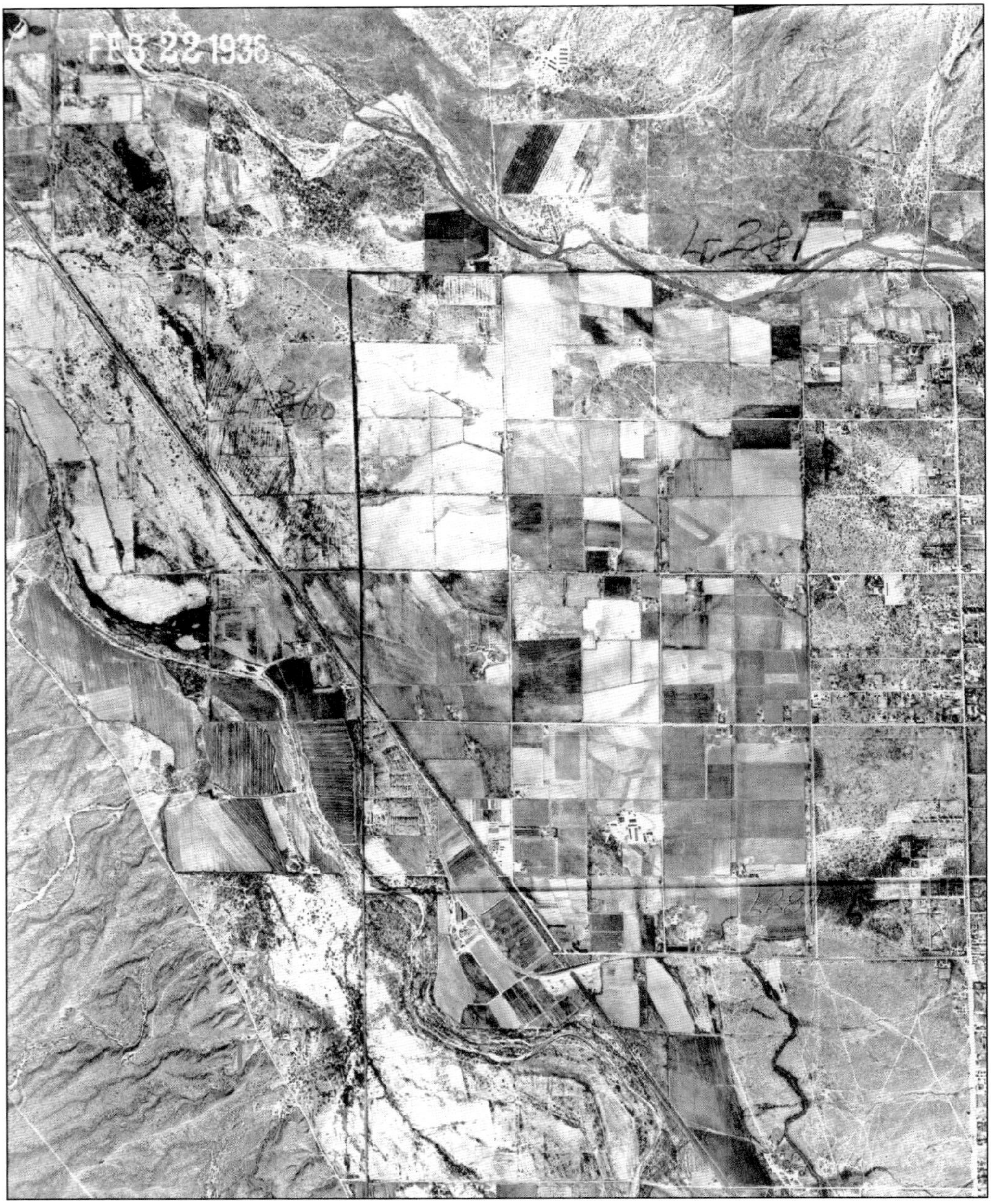

This 1936 aerial view of the Flowing Wells area prominently features railroad tracks and the Flowing Wells Canal entering the area at bottom right. At the top is the Rillito River. At left is the Santa Cruz River. Fairview Road is on the right, just to the right of the Flowing Wells Canal. Near bottom center, the transient camp can be seen just off Flowing Wells Road. The railroad came to Tucson in 1880. (Pima County Government Maps.)

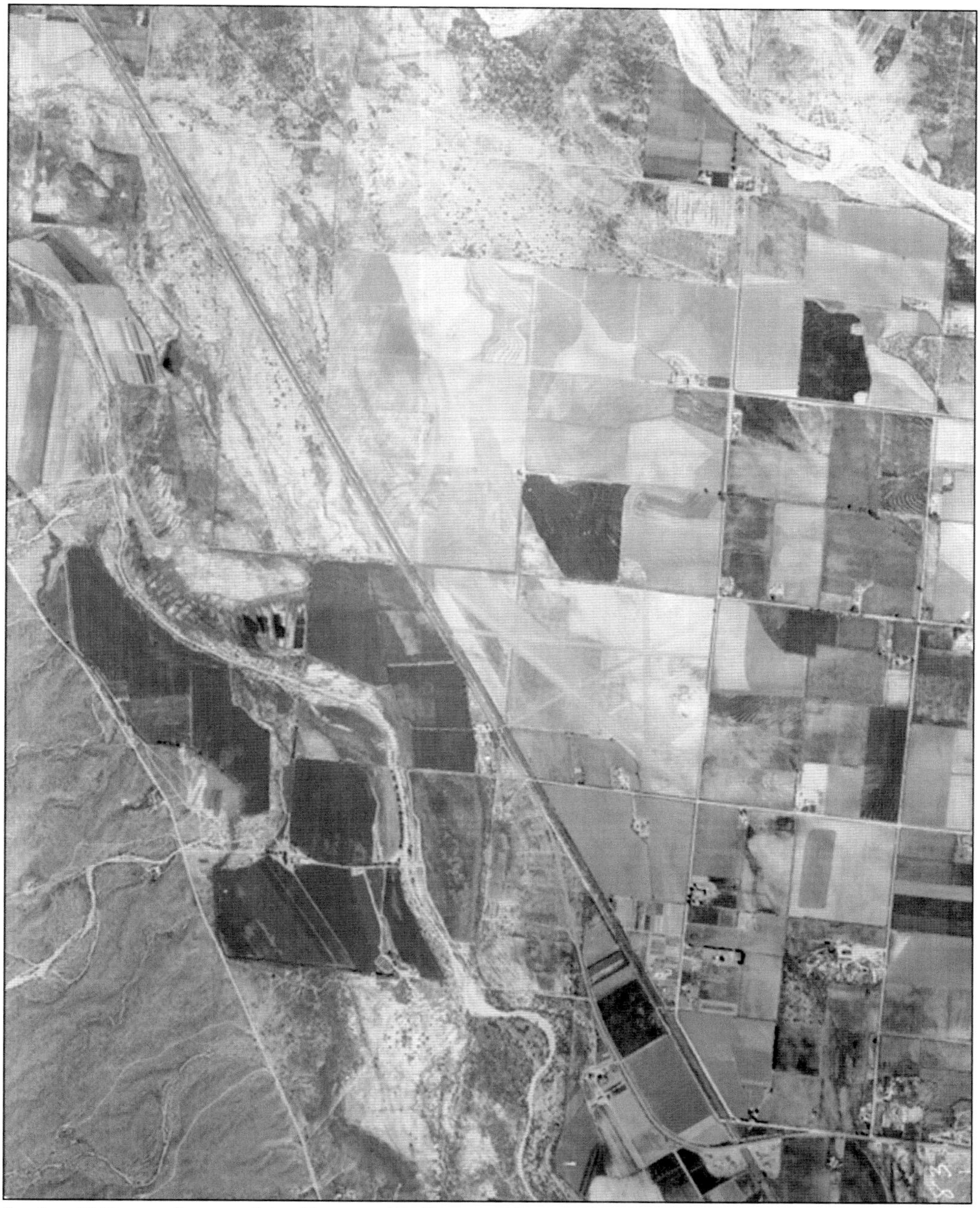

In this 1941 aerial view of the Flowing Wells area, Shamrock Dairy can be seen at top right center on the bank of the Rillito River. The new Gilpin Airport's dirt runways are clearly seen at Prince and Romero Roads. (Pima County Government Maps.)

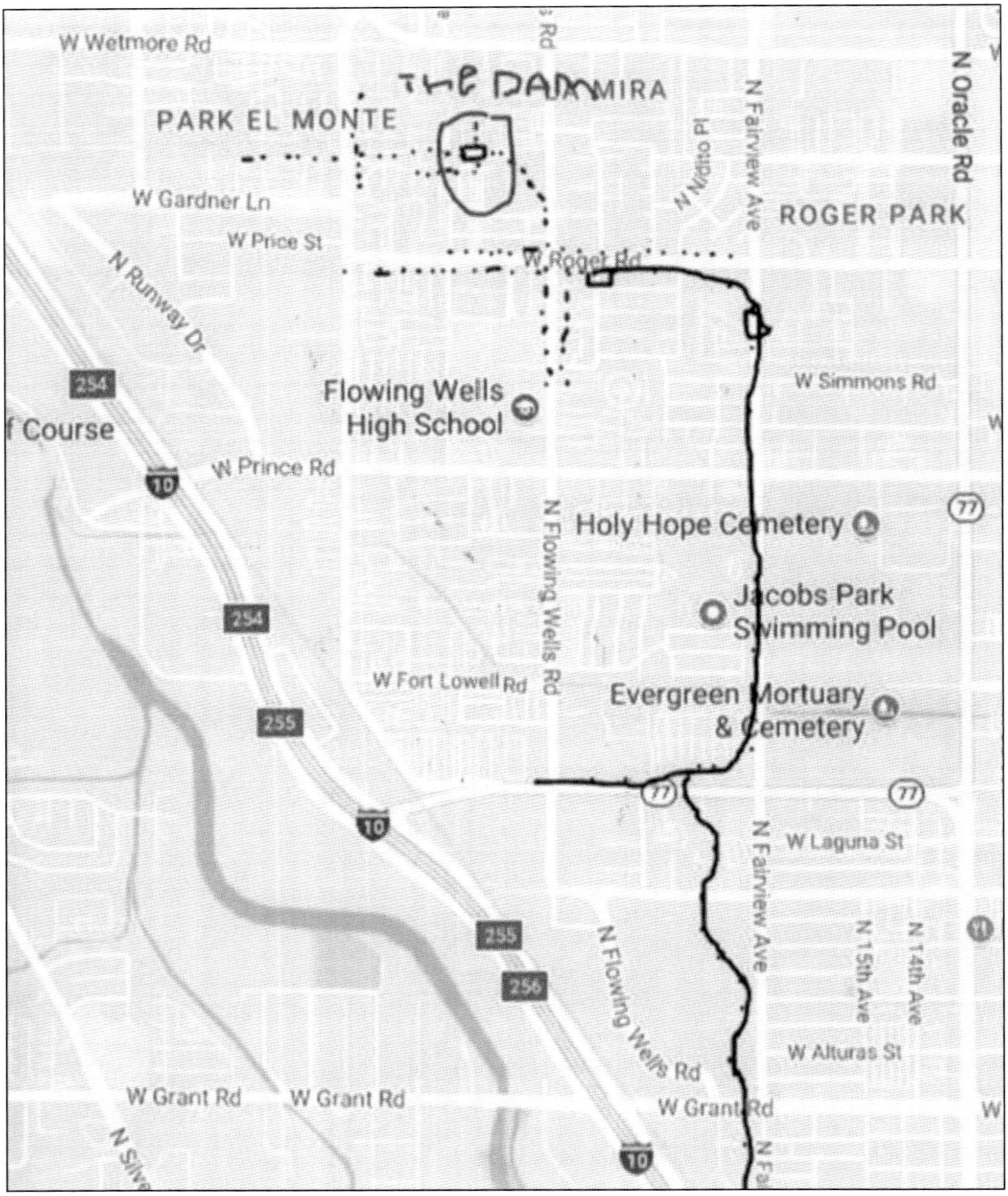

This is a sketch of the larger canals in the Flowing Wells area. These canals were filled by pumps (the square areas along the route). There was a significant area where kids would swim in the 1950s called "the Dam." Several smaller irrigation ditches flowed from this structure. The canal map is overlaid on a current map of the area. (James Elliot.)

Two

Early Settling

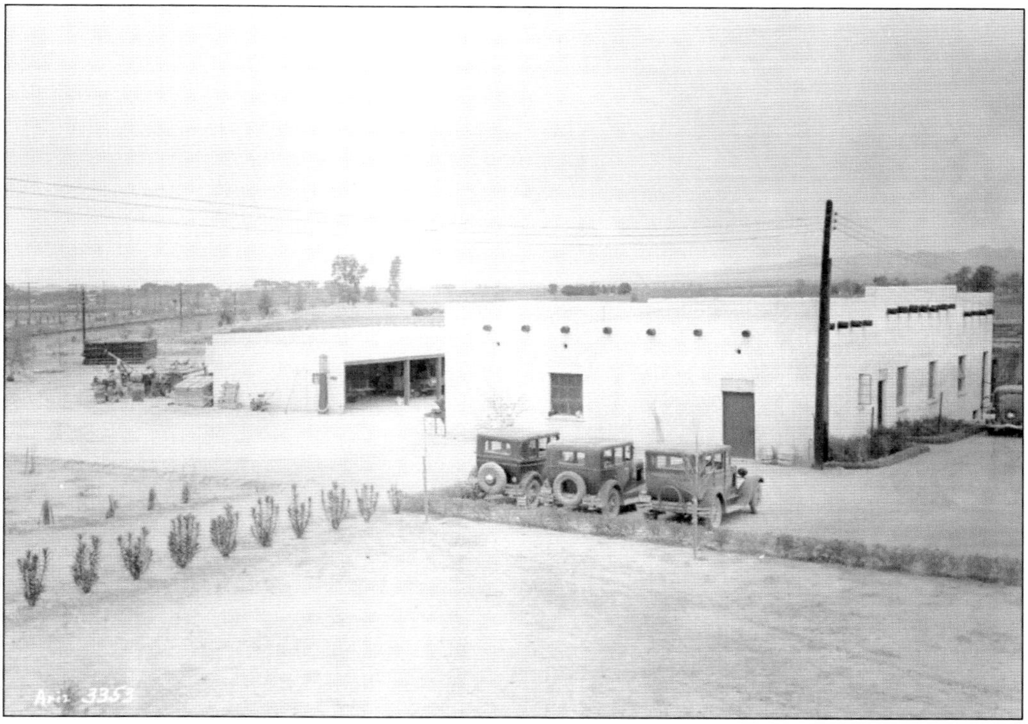

The Tucson Plant Materials Center was constructed in 1934 using local labor and funded by the Federal Emergency Relief Act. It sits on approximately 18 acres on Romero and Fort Lowell Roads (formerly Sunshine Lane). Originally called an erosion nursery, the facility was built by the Bureau of Plant Industry in response to the devastation of the Dust Bowl era of the 1930s. Railroad tracks can be seen in the left background, and the Tortolita Mountains in the right background. Dodge sedans are parked in front. (USDA NRCS.)

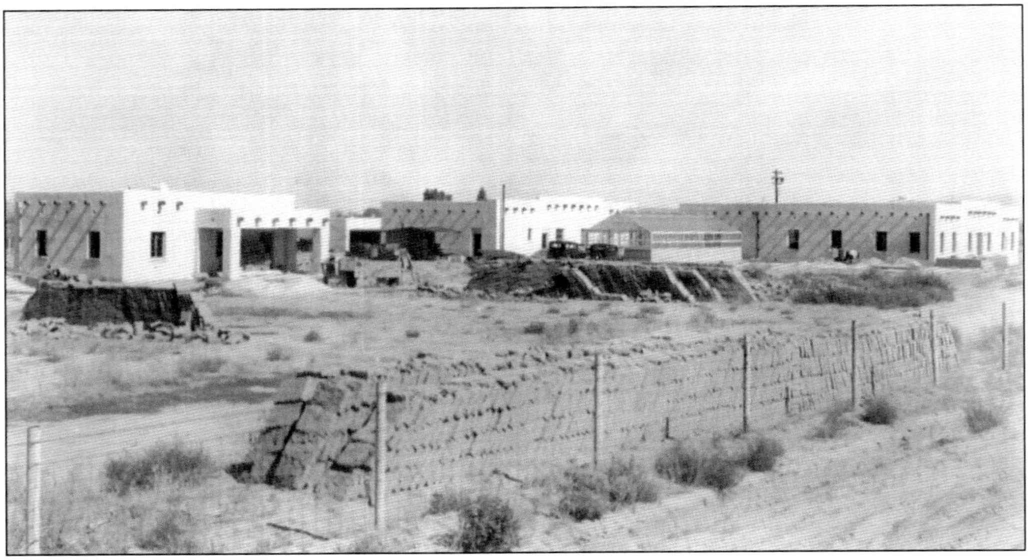

The Tucson Plant Materials Center was constructed of adobe bricks made by local labor. Adobe, meaning "mud brick" in Spanish, is made up of earth, other organic material such as straw, and water. The bricks are sun-dried in forms, then stacked together to build structures. Adobe is among the earliest building materials used in Tucson. (USDA NRCS.)

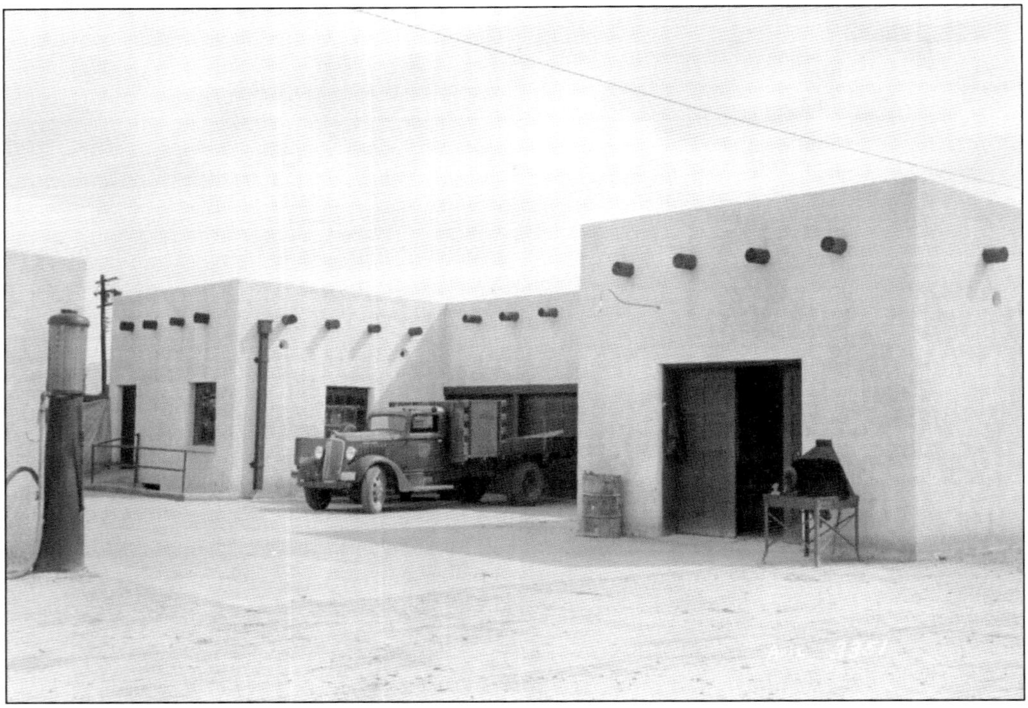

The purpose of the Tucson Plant Materials Center was to propagate and observe plants for use in reducing soil erosion, produce large quantities of plants for use in the Gila, Navajo, and Rio Grande regions, and determine techniques to enhance plant production. Currently, the Tucson Plant Materials Center maintains 11 plant species for conservation and use and has another three species in initial evaluation for release. This is one of the original adobe buildings constructed in 1934, with a Studebaker truck and an old-fashioned gas pump. (USDA NRCS.)

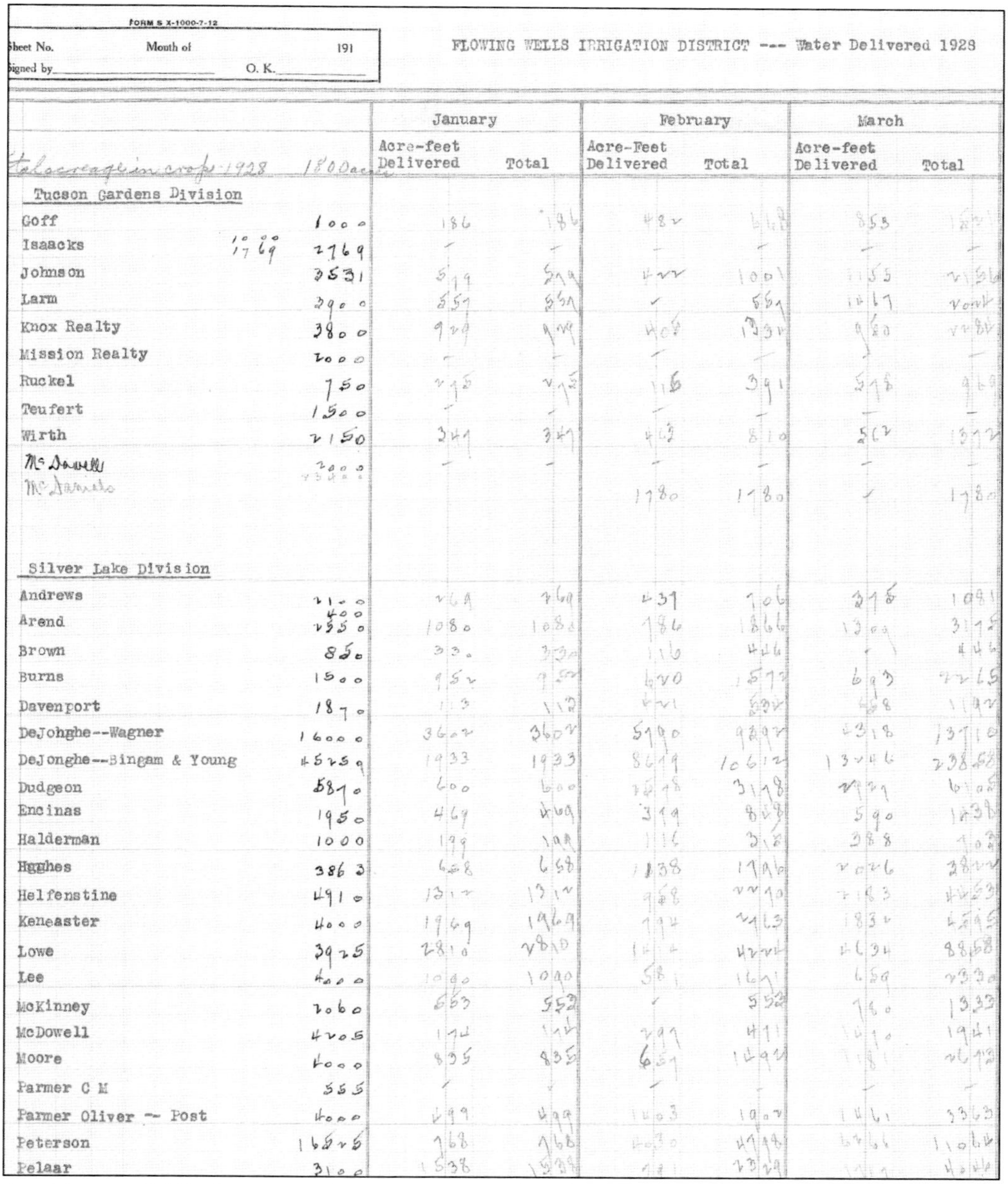

FORM S X-1000-7-12

Sheet No. Month of 191

Signed by O. K.

FLOWING WELLS IRRIGATION DISTRICT --- Water Delivered 1928

Total acreage in crop 1928 1800 acres		January Acre-feet Delivered	January Total	February Acre-Feet Delivered	February Total	March Acre-feet Delivered	March Total
Tucson Gardens Division							
Goff	1000	186	186	482	668	853	[illegible]
Isaacks 1000 1769	2769	–	–	–	–	–	–
Johnson	3531	519	519	422	1001	1155	2156
Larm	2900	557	557	✓	557	1467	[illegible]
Knox Realty	3800	929	929	402	1331	980	[illegible]
Mission Realty	2000	–	–	–	–	–	–
Ruckel	750	275	275	116	391	578	969
Teufert	1500	–	–	–	–	–	–
Wirth	2150	347	347	463	810	562	1372
McDowell	2000	–	–	–	–	–	–
McDaniels	2340			1780	1780	✓	1780
Silver Lake Division							
Andrews	2100	269	269	437	706	375	1081
Arend	400 2550	1080	1080	786	1866	1309	3175
Brown	850	330	330	116	446	–	446
Burns	1500	952	952	620	1572	693	2265
Davenport	1870	113	113	421	534	658	1192
DeJohghe--Wagner	16000	3602	3602	5990	9592	4318	13710
DeJonghe--Bingam & Young	45259	1933	1933	8619	10612	13246	23858
Dudgeon	5870	600	600	[illegible]	3178	[illegible]	[illegible]
Encinas	1950	469	469	379	848	590	1438
Halderman	1000	199	199	116	315	388	703
Hughes	3863	658	658	1338	1996	2026	3822
Helfenstine	4910	1312	1312	978	2290	2183	[illegible]
Keneaster	4000	1969	1969	194	2163	1832	[illegible]
Lowe	3925	2810	2810	[illegible]	4224	4634	8858
Lee	4000	1090	1090	581	1671	659	2330
McKinney	2060	553	552	✓	552	780	1333
McDowell	4205	122	122	289	411	1410	1941
Moore	4000	835	835	657	1492	1181	2673
Parmer C M	555	✓	✓	–	–	–	–
Parmer Oliver -- Post	4000	499	499	1403	1902	1461	3363
Peterson	16525	768	768	4030	4798	6266	11064
Pelaar	3100	1538	1538	[illegible]	2329	[illegible]	[illegible]

The Flowing Wells Irrigation District was created on May 2, 1922, by the Pima County Board of Supervisors along with the Jaynes Irrigation District. Thirty-four farmers proposed that they could produce water better and cheaper than the Tucson Farms Company. The Great Depression was hard on FWID. Paul M. Arend (whose home was at the northwest corner of Miracle Mile and Fairview Avenue), William T. Dudgeon, John C. Morgan, James G. Keneaster, and others worked diligently and successfully prevented bankruptcy, resulting in good health by 1936. The Reconstruction Finance Corporation was created in 1933 by the US government because it was not in the public interest to have farms deserted and unproductive. Overall, the Flowing Wells Irrigation District was looked upon favorably as one of the best and clearest irrigation projects as a candidate for a loan to get through the Depression. With help from Arend, who mortgaged some of his property to meet the obligations for the federal loan, Sen. Carl Hayden, and Congresswoman Isabella Greenway, FWID received a $44,500 federal loan in 1935. (FWID.)

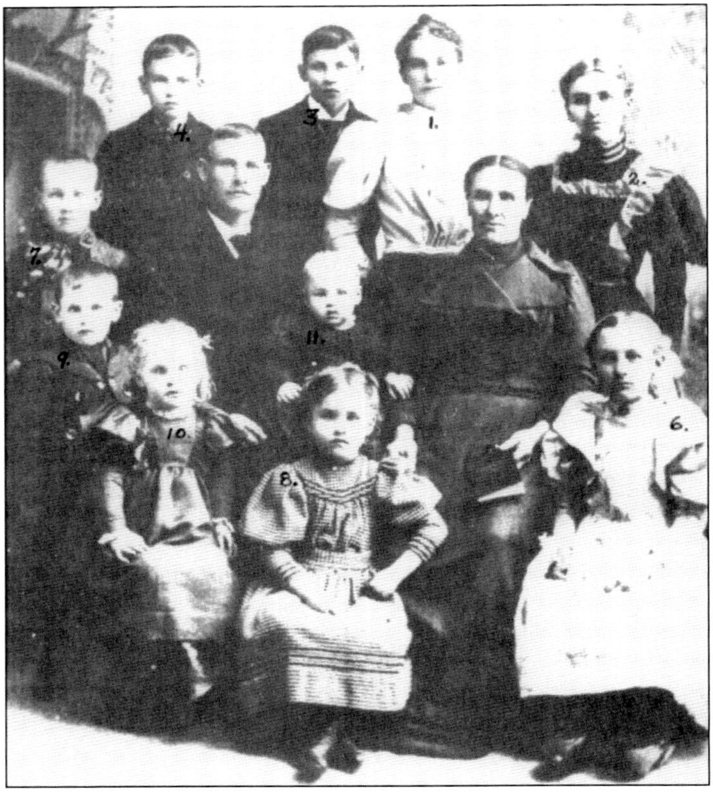

The Fredrik and Merit Loset Peterson family is pictured around 1897. Henry (top left) started the Holstein (Peterson Bros.) Dairy in 1904 in Tucson. Also known as Autry & Peterson, it is believed to be the first modern dairy in Tucson. Henry's younger brother John (far left) joined him in the dairy business. Hjalmar, the youngest (center), came to Tucson in 1924 at the age of 28. He met Margaret Nagel, and they married on December 26, 1924, raising two sons, Ronald and Maynard "Pete" H. Peterson. Pete is a lifelong Daily family friend and is the stepfather of coauthor Kevin Daily. (Maynard "Pete" H. and Patricia Peterson.)

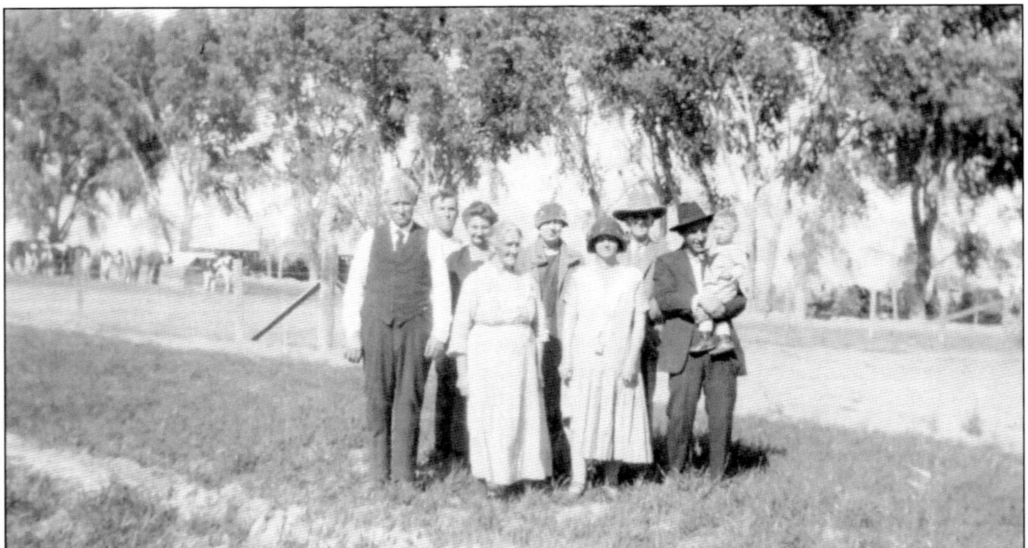

The Petersons are pictured around 1924. From left to right are Fredrik, Henry, Mamie Bouschet Peterson, Marit, possibly Anna Peterson Ryning, Cora Benson Peterson, possibly Henry Ryning, John, and child Duane. Henry Peterson served as one of the three members of the first Flowing Wells School Board, serving two terms from 1928 to 1935, the year of his death. This photograph was taken at John and Cora's home south of Sentinel Peak on Mission Road. (Maynard "Pete" H. and Patricia Peterson.)

JAYNES STATION

FLOWING WELLS

Frank Stock of Flowing Wells, who enlisted in the medical department last July, returned home on a two-day furlough Sunday morning. In honor of his return his parents gave a Christmas dinner Monday. A dance also was given in his honor Monday evening at the Davidson school house. He returned to Camp Fremont, California, Tuesday morning.

Miss Sprague, who teaches the school at Flowing Wells, and her mother, are visiting at Flowing Wells until the opening of school.

Ben-A. Johnson of the Tucson Farms company left last week for his home in Denver.

J.P.Bracketthasresignedhis ....

J. P. Brockett has resigned his position as bookkeeper for the Flowing Wells dairy and has returned to his home in Denver.

Mr. Sunderland has accepted the position a sbookkeeper for the Flowing Wells dairy.

E. W. Taylor, general manager of the Tucson Farms company, is in California on business.

Both the Jaynes and Flowing Wells Sunday schools held their Christmas exercises Christmas eve. The Jaynes Sunday school met at the Jaynes school house, whiel the Flowing Wells school met at the Flowing Wells school house. Both communities enjoyed a tree and each one present was presented a Christmas bag.

Mr. Hilger, former bookkeeper for the Flowing Wells dairy, is seriously ill at the Arizona hospital.

Miss Lillian Watcher of Jaynes has been quite ill since Christmas.

K. W. Stubblefield of Flowing Wells is still confined to his bed.

Mr. and Mrs. Sinclair of Jaynes entertained Mr. and Mrs. Rose at dinner on Christmas day.

Mr. and Mrs. Matt Watcher and family entertained Mr. and Mrs. Gerhart and family at dinner on Christmas.

Professor Medcraft of the University of Arizona is installing a new engine on his ranch at Jaynes Station.

Mrs. Bard, mother of Mrs. Gressinger, who has been ill for a long time, is still confined to her bed.

Mr. and Mrs. C. A. Traver and family of Flowing Wells entertained Willis Walker of Jaynes Station at dinner on Christmas.

A farewell party was given at the home of Mr. and Mrs. Wood at Flowing Wells in honor of J. P. Brackett and Ben A. Johnson. The evening was spent in playing progressive pedro. The guests included Miss Sprague, Miss Ginn, Mrs. Hiss, Mrs. Gotthelf, Mrs. Hopely, Lucy Hopely, Mrs Perry, Mr. Sunderland, Joe Hiss, J. P. Brackett, Ben A. Johnson, Adolph and Gus Petrie and Mr. Hopely.

Flowing Wells and Jaynes Station have always been tight-knit. Flowing Wells Sunday school was started in 1916 and by 1918 had 47 regular members. Jean Gotthelf directed the singing and Nancy Pelaar was pianist, with Helen Wetmore and Llynn Hodgeson as violinists. Sunday school was held at the Flowing Wells Schoolhouse on Romero Road. Mrs. Wetmore and Nancy Pelaar formed the Sunday school. Jaynes Station had its own post office from 1922 to 1924. Marcia Gardner Claridge remembers a time when the outgoing mail was placed on a hook for train workers to grab and the incoming mail bag was thrown to the ground. Jaynes Station residents often flagged down trains for rides into Tucson. The Oracle–Jaynes Station Road connected Oracle Road and ended at Jaynes Station. (*Arizona Daily Star.*)

Jirou Enters Plea of Guilty to Part In Train Hold-Up

Imposition of Sentence Is Indefinitely Suspended by Judge Pattee

Imposition of sentence was indefinitely suspended on Frank Jirou, who yesterday pleaded guilty before Judge Samuel L. Pattee in superior court to a charge of assisting in the hold-up of the Golden State Limited, 8 miles from Tucson last May.

The motion to suspend sentence on Jirou was made by the defense counsel, Howard C. Griffin. It was stated by County Attorney George R. Darnell that he believed the defendant would not be sentenced until after the trial of the other men alleged to have been implicated in the attempted train robbery.

Jirou, in a confession at his preliminary examination, implicated George Winkler and his two sons, Edward and George Jr., Tom Dugat, who was killed, and a Mexican goat herder, Santiago Valdez. Edward Winkler and the goat herder are still at large.

The case of George Winkler Jr., was certified to juvenile court when it was learned that the youth was under age.

Jaynes Station was the location of the last attempted Arizona train robbery in 1922. In a failed attempt to rob the *Golden State Limited* on its way into Jaynes Station on May 15, 1922, goat farmer Tom Dugat, who had fallen on hard times, was killed by a load of buckshot fired by railroad employee Harry Stewart. Three masked men boarded the train, while two others held positions along the track. Stewart was watching the events unfold and picked the right moment. The four remaining would-be robbers at first got away. Some sped away in the getaway car, and some scattered into the desert, but all were apprehended within weeks except one who fled to Mexico. Frank Jirou turned state's evidence, snitching on the rest of the gang. Stewart was presented with an engraved watch, a $1,000 check, and a letter of appreciation from the railway company. (*Arizona Daily Star* and *Arizona Republic*.)

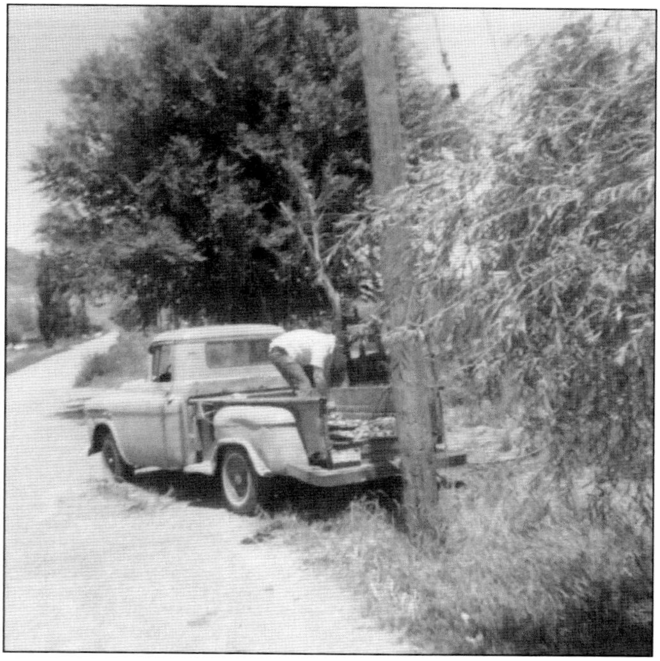

Brothers Mark, Quince, and Taylor Gardner bought 20 acres on Romero Road from E.M. Furrey, son of G.L. Furrey. They built their houses here, initially without electricity. Roger Gardner (FWHS 1969) is shown unloading apples from a 1965 Chevrolet pickup on the corner of Gardner Lane and Romero Road. Mark Gardner donated the land to the county to create the road now known as Gardner Lane. (Marcia Gardner Claridge.)

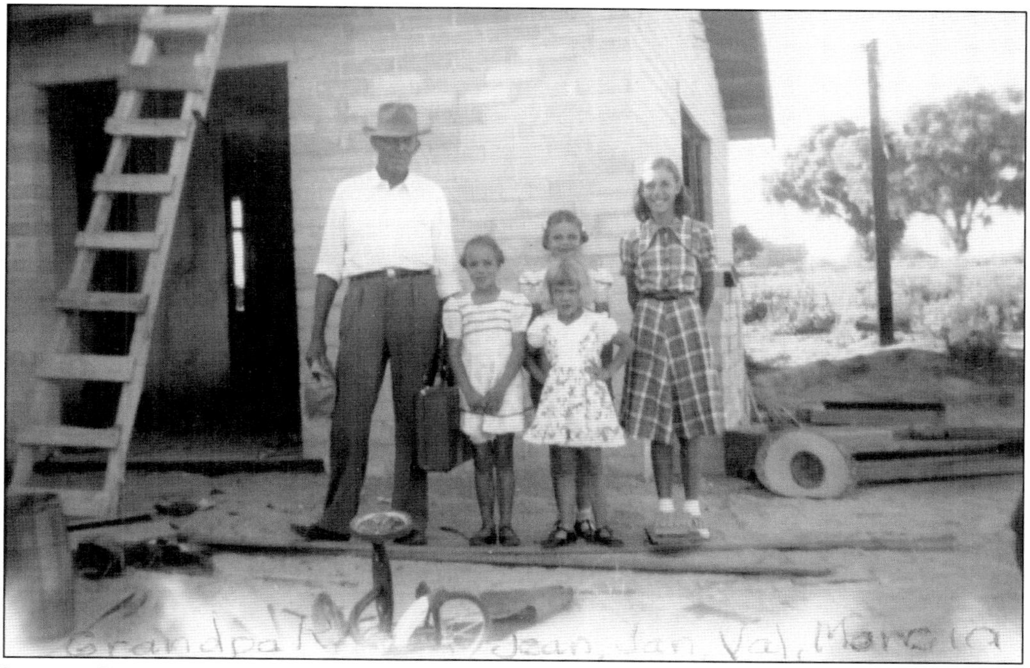

Lorenzo Rogers is pictured with his granddaughters, the daughters of Mark and Inez Gardner. From left to right are Jean and Jan (FWHS 1959), Val (FWHS 1960), and Marcia (Tucson High School). They are pictured in front of their house, under construction in 1949. Mark served on the FWID board for approximately 25 years. Inez taught fourth grade at Laguna and Homer Davis Elementary Schools from around 1955 until her retirement in 1976. Marcia was in the first full ninth-grade class at Flowing Wells in 1952. She married Dr. Roy Claridge, who taught and coached at FWHS from 1960 to 1967. They raised six kids, all attending Flowing Wells schools. (Marcia Gardner Claridge.)

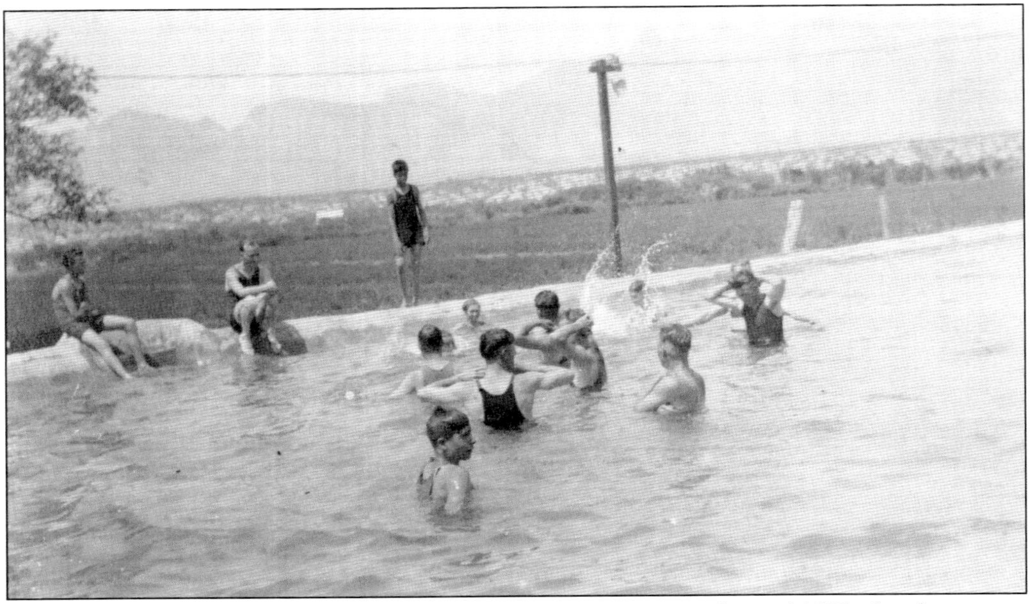

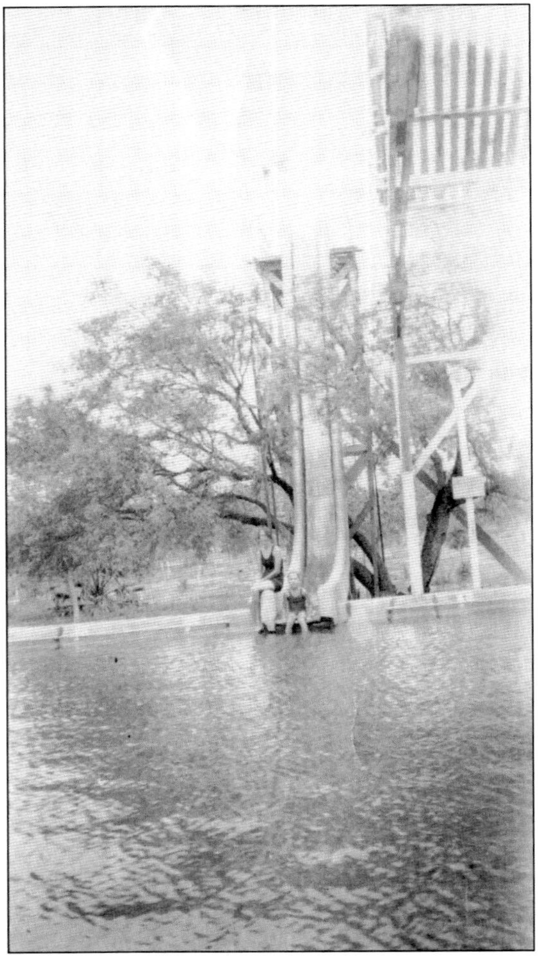

From 1918 to the mid-1970s, locals swam in the Wetmore Pool. Helen Wetmore used to swim in the irrigation ditch near their property until her uncle Ed built her a 15-by-30-foot pool. The popularity of her small pool gave her brother-in-law the idea to build a larger one that winter, using horses with scoops to dig out the dirt and laying all the concrete by hand. It was the state's first Olympic-size pool. With no filtration system, the water was completely drained out twice a week to irrigate the surrounding trees. (Arizona Historical Society.)

The diving board was originally 30 feet high but was later lowered to 18 feet. Copper sulfate was used to keep the pool clean. It was reported that many local children had green hair from the chemical. After World War I, brothers Edward and Ralph Wetmore graded the road along their property using a team of horses. This is now Wetmore Road. Helen Wetmore had the idea for the Tucson Mall, and in 1978, her thoughts became a reality when the mall was built on Wetmore family land. (Arizona Historical Society.)

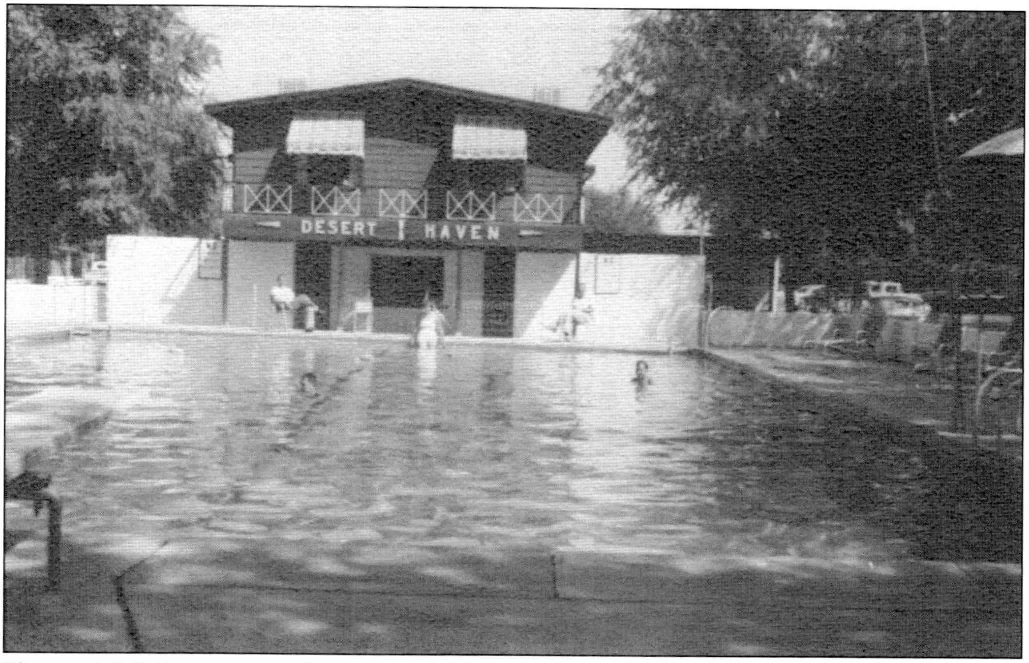

Flowing Wells had another favorite pool at Desert Haven Mobile Home Park on Romero Road, between Sunshine Lane and Prince Road. Swimming pools were not so popular during the polio scare in the 1950s. Marcia Gardner Claridge remembers standing in line at the FWHS cafeteria to get the polio vaccination on a sugar cube. (FWUSD.)

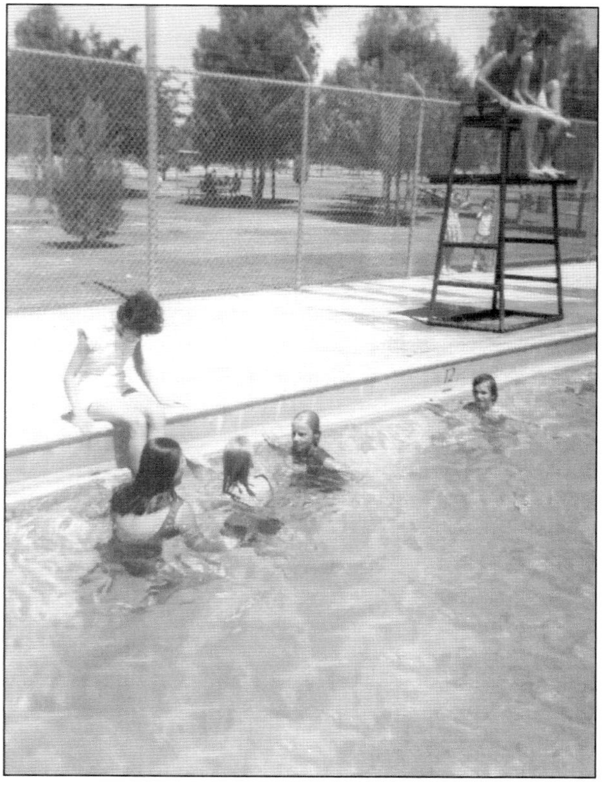

In 1959, Flowing Wells was annexed into Tucson and was promised the Del Norte Park Pool (now Jacobs Park Pool), opened in 1963 on Fairview Lane and Mohave Road. Gene C. Reid was the director of parks and recreation at the time. In 1972, Del Norte Park was renamed for one of Tucson's most respected coaches, Amphitheater High School football coach (starting 1960), educator, and award-winning football player Bob Jacobs, who was killed in a 1971 plane crash. Jacobs's sons Mark (FWHS 1973), Mike (FWHS 1974), and Rory all attended FWHS. Mark and Mike were award-winning football players. Mark was a linebacker on the 1972 All-State team and was inducted into the FWHS Athletics Hall of Fame's first class in 2003. Mike (a defensive halfback) and Mark both won the *Arizona Daily Star* Tom McAn Shoe Company Award for outstanding football players in Tucson. (FWUSD.)

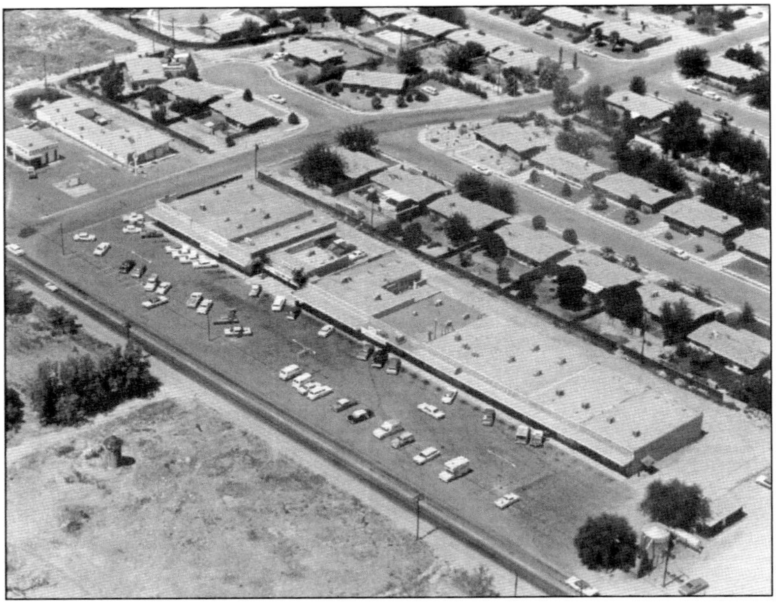

This c. 1960 aerial view shows the shopping center at Fairview Avenue and Prince Road. The Flowing Wells Irrigation District office can be seen at lower right. Lowry Hardware moved to the right end of this shopping center from Stone Avenue. The corner was previously the location of the William T. Dudgeon family home. (Eddie Lowry and Sue Lowry Rhude.)

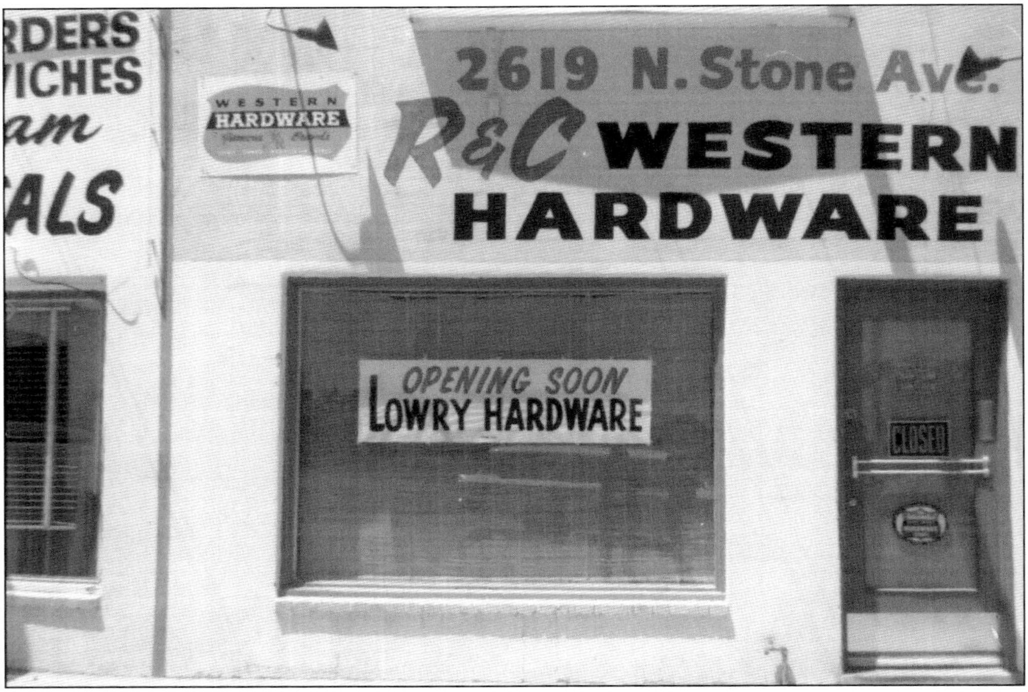

Lowry Hardware originally opened on Stone Avenue. Moving to Fairview Avenue and Prince Road in the early 1960s, it has been a staple in the community for over 55 years. It is known to locals as the go-to place for just about anything they could ever need from a hardware store. Lowry's carries smaller and harder-to-find goods and caters to the average homeowner. Customer service has always been a top priority at Lowry's. Since 1980, Lowry Hardware has been located at 975 West Prince Road. (Eddie Lowry and Sue Lowry Rhude.)

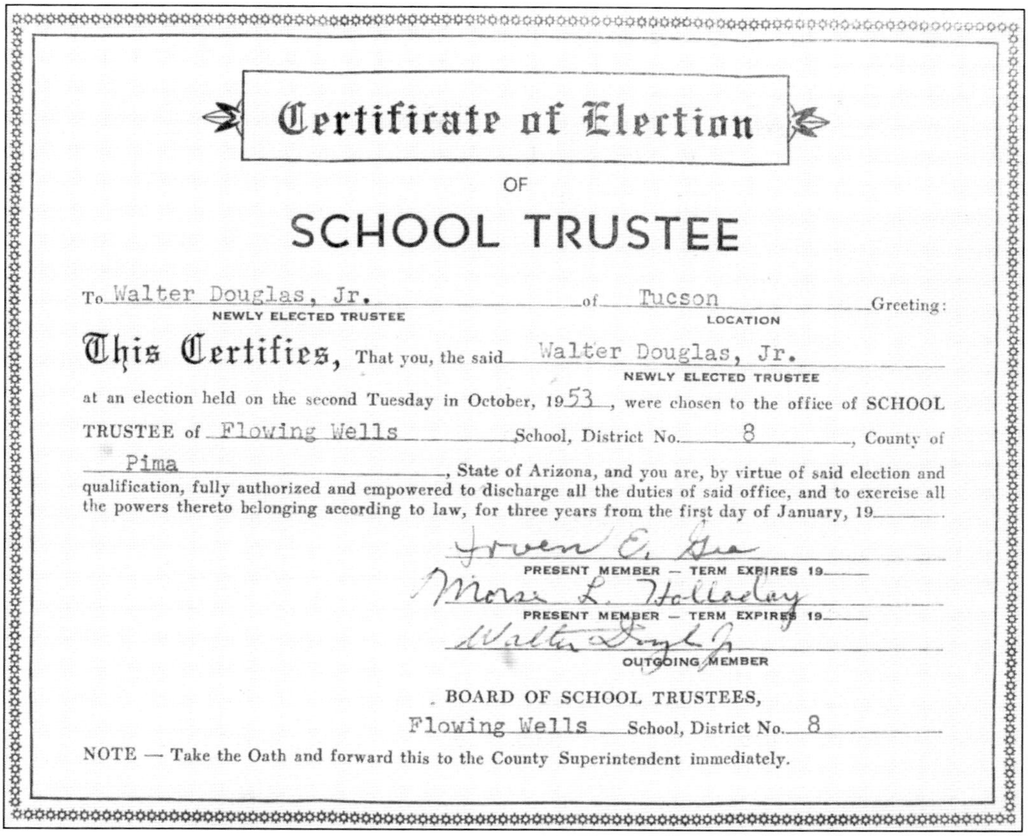

Certificate of Election

OF

SCHOOL TRUSTEE

To Walter Douglas, Jr. of Tucson Greeting:
NEWLY ELECTED TRUSTEE — LOCATION

This Certifies, That you, the said Walter Douglas, Jr.
NEWLY ELECTED TRUSTEE

at an election held on the second Tuesday in October, 1953, were chosen to the office of SCHOOL TRUSTEE of Flowing Wells School, District No. 8, County of Pima, State of Arizona, and you are, by virtue of said election and qualification, fully authorized and empowered to discharge all the duties of said office, and to exercise all the powers thereto belonging according to law, for three years from the first day of January, 19___.

Irven E. Gee
PRESENT MEMBER — TERM EXPIRES 19___

Morse L. Holladay
PRESENT MEMBER — TERM EXPIRES 19___

Walter Doyle Jr.
OUTGOING MEMBER

BOARD OF SCHOOL TRUSTEES,
Flowing Wells School, District No. 8

NOTE — Take the Oath and forward this to the County Superintendent immediately.

Walter Douglas Jr. was first appointed to the Flowing Wells School Board in 1942. At the time, there were six teachers and 70 students at the school. When his term ended, he ran for reelection. As the proprietor of Gilpin Airlines, he was the second-largest taxpayer in the district, and said he "didn't want to see his money wasted on frivolous programs." In 1982, Douglas received a commendation from Pres. Ronald Reagan for being the longest-serving school board president. He remained on the school board for 45 consecutive years until his death in 1987. (FWUSD.)

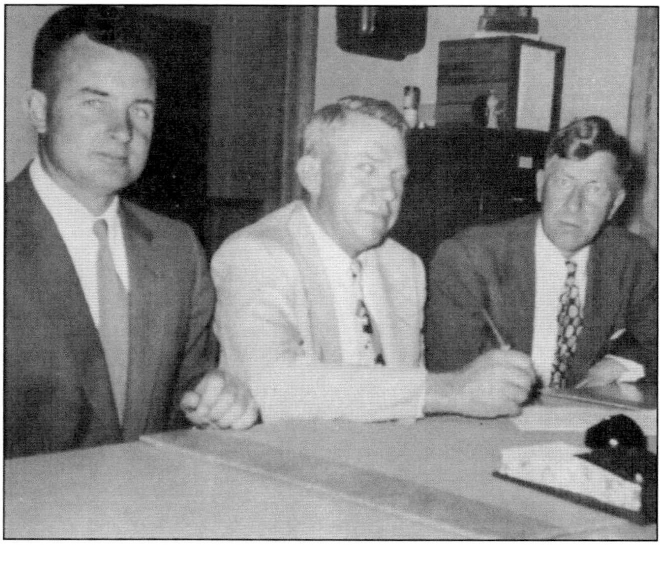

The 1954 Flowing Wells School Board consisted of, from left to right, Morse Holladay, Irven Gee, and Walter Douglas, three local businessmen and fathers of local kids. On January 15, 1956, Irven Gee was inspecting the progress of construction at the new FWHS and witnessed the crash of a private 1950 Beechcraft Bonanza at what is now approximately the finish line of the high school track. In June 1973, there was another crash landing on one of FWHS's athletic fields. (FWUSD.)

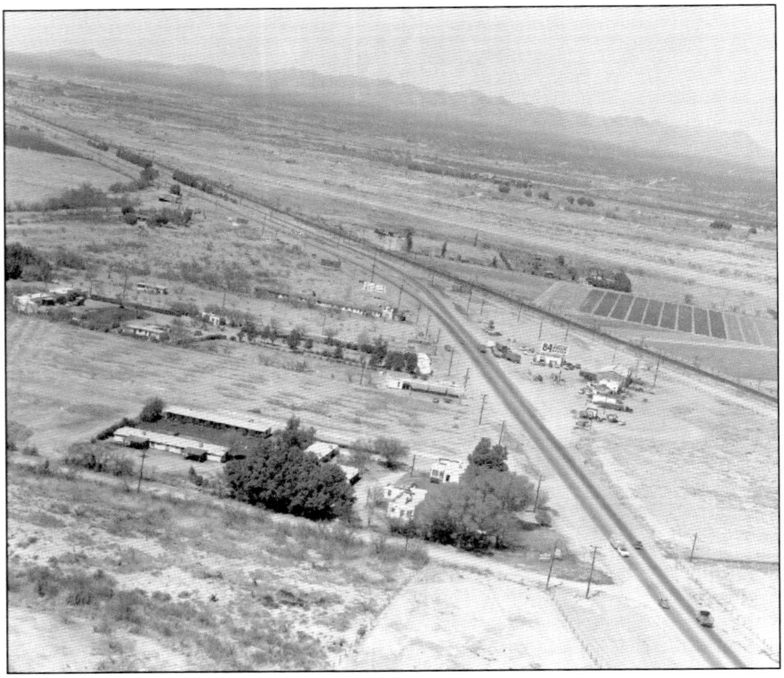

This c. 1950 aerial view was taken facing north with the 84 Truck Stop at right center and the Tucson Plant Materials Center land south of Prince Road and the old Casa Grande Highway (State Route 84). Interstate 10 had not yet been constructed. (Arizona Historical Society.)

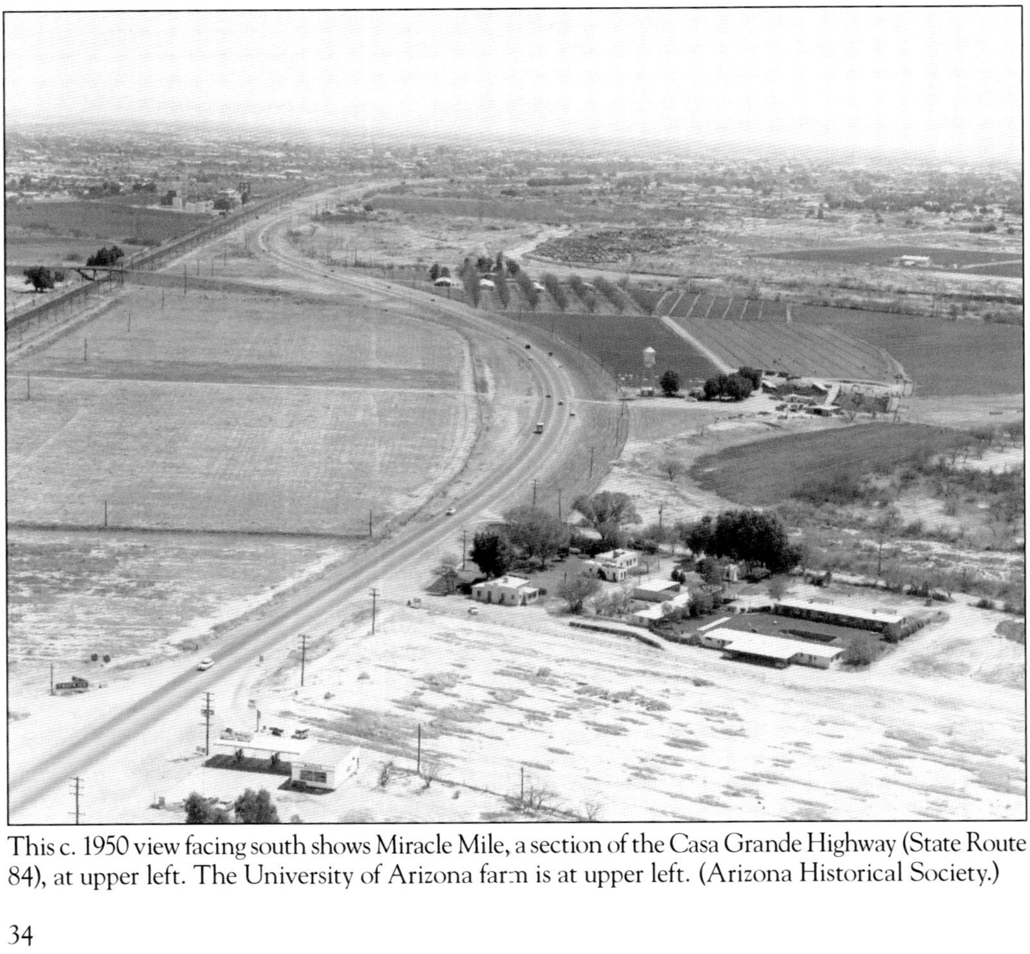

This c. 1950 view facing south shows Miracle Mile, a section of the Casa Grande Highway (State Route 84), at upper left. The University of Arizona farm is at upper left. (Arizona Historical Society.)

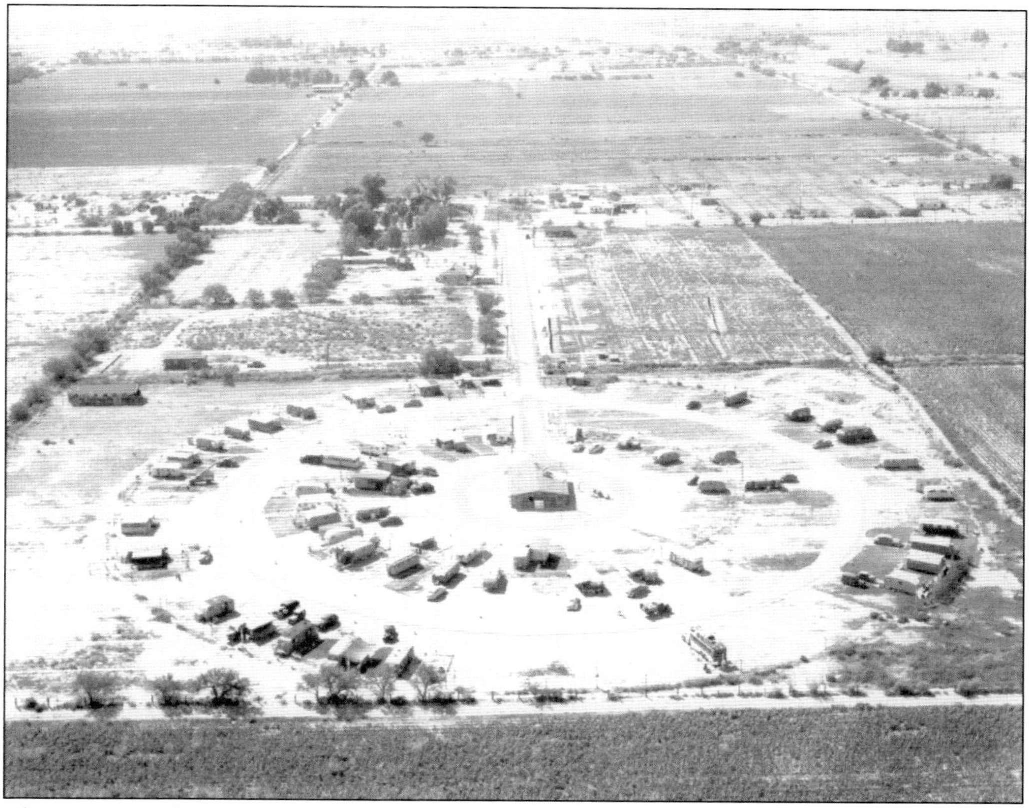

This 1947 aerial view shows Crescent Manor Mobile Home Park with its distinctive circular layout. The view is to the south. The main road in the center of the circle ends at Prince Road. In 1951, Crescent Manor offered free bus transportation to Grand Central Aircraft and Hughes Aircraft for those who lived in the park. (Arizona Historical Society.)

CRESCENT MANOR MOBILE HOME PARK

Fifteen acres of happy living. Shopping center on premises. Natural gas and 220 voltage. 10 minutes from town. Laundry - city sewers - shady lots - individual clothes lines - shuffleboards - large recreation hall.

Hy & Molly Shenitzer, Owners

1150 W. Prince Rd. Tucson, Arizona
MA 2-4100

Crescent Manor offered all the latest amenities of its time, including individual clothes lines. Just one block from the Flowing Wells School, the location was convenient for families. Crescent Manor advertised that it had a generator in case the power went out. (FWUSD.)

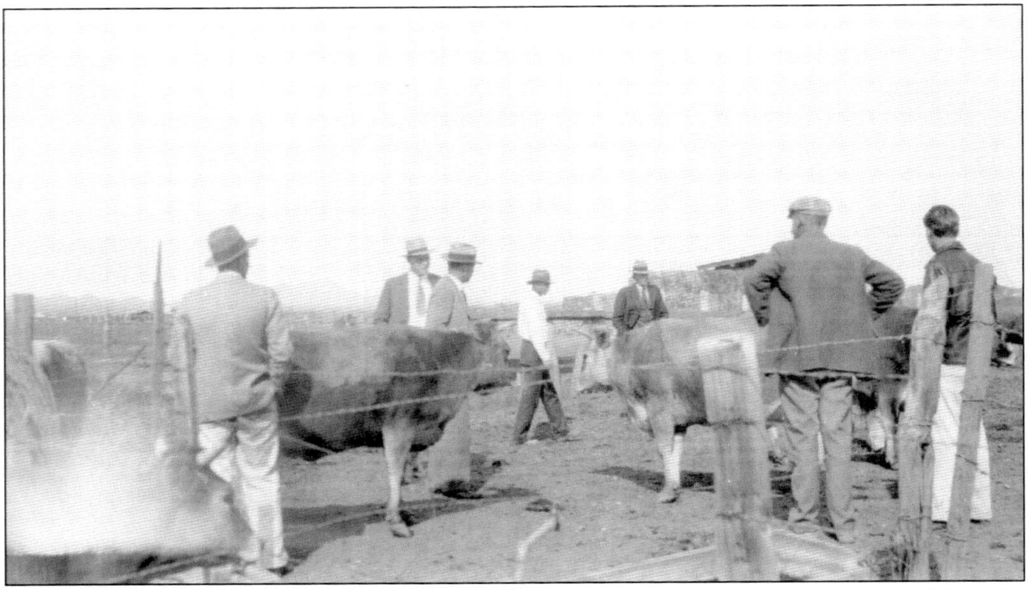

William Thomas Dudgeon (1879–1973) was a dairyman, agriculture experimenter, and founder of the Catalina and Menlo Park Methodist Churches. Dudgeon and his first wife, Sardina Ford, came to Tucson in 1922. Dudgeon's home and 40 acres were on the northwest corner of Prince Road and Fairview Avenue. In 1925, he purchased 20 head of purebred cattle to start a dairy. Shamrock and Sunset Dairies bought milk from him. He served as president of the Milk Producers Association around 1925. This cow judging demonstration took place at the Dudgeon farm on March 5, 1931. After Sardina passed, Dudgeon and his second wife, Iola, built a home at the orchard on Roger Road in the 1950s. (Arizona Historical Society.)

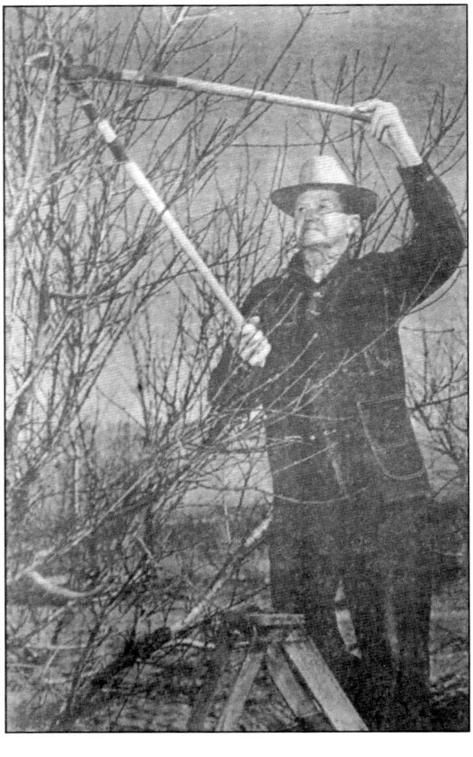

W.T. Dudgeon is pictured in 1952. In 1948, he began an experimental 10-acre fruit orchard at 1104 West Roger Road. Willola Orchard was set up to prove that Tucson's climate was good for fruit production. The orchard became a thriving producer of apricots, plums, and peaches and was used to develop pruning methods in conjunction with the University of Arizona's College of Agriculture. From 1929 to 1935, Dudgeon helped reorganize the Flowing Wells Irrigation District. Both he and his wife Sardina served on the Flowing Wells School Board individually between 1922 and 1938. (*Tucson Daily Citizen* and the University of Arizona Special Collections.)

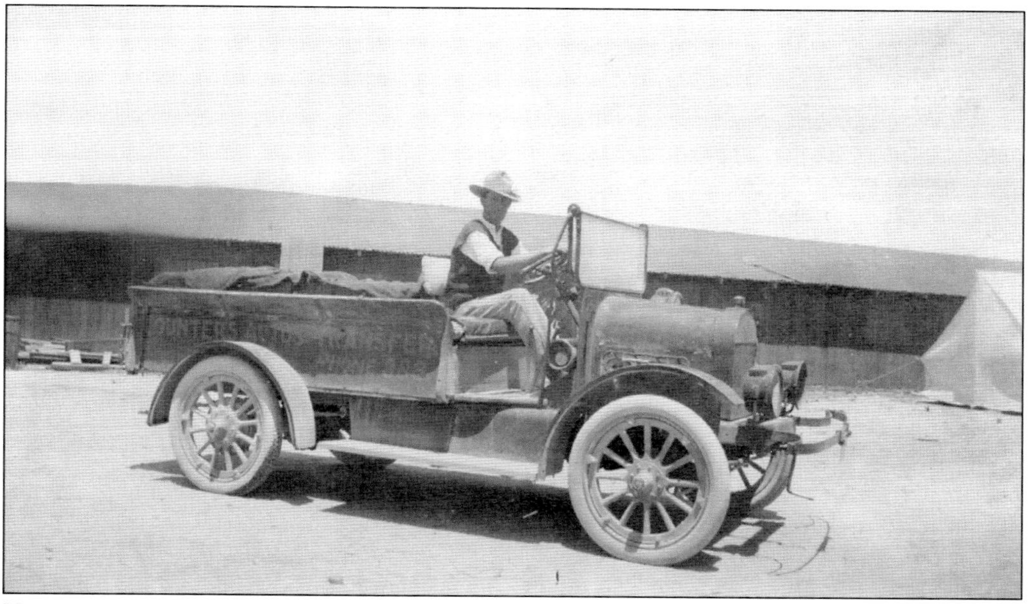

Henry Peterson started one of the first modern dairies in Tucson in 1904. The Holstein Dairy, also named the Peterson Dairy, had one of its farms on Prince Road near what is now Interstate 10. The truck in this photograph is a 1920s International. (Diane Peterson Graham and Maynard "Pete" H. Peterson.)

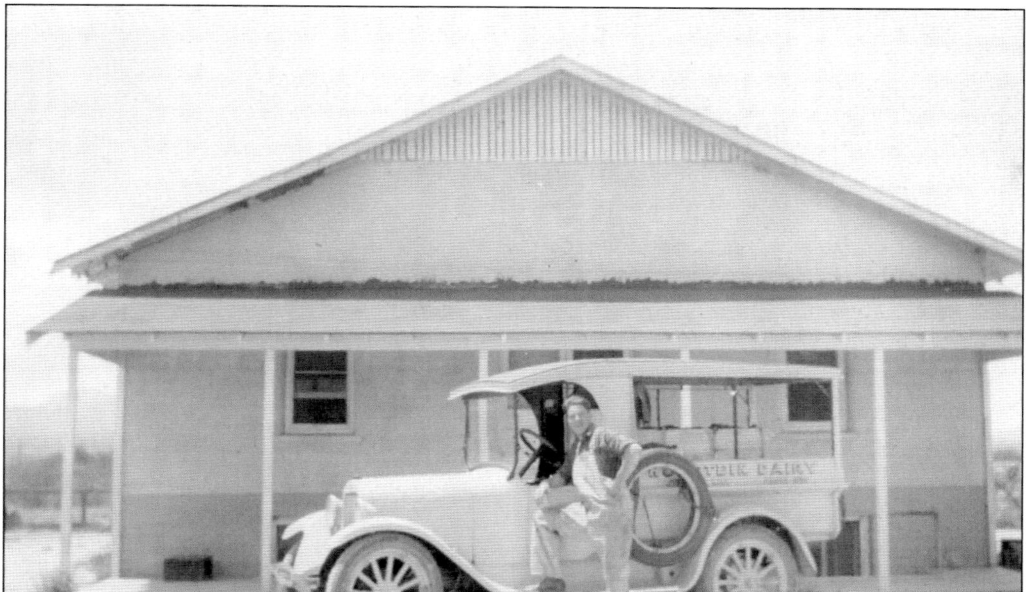

Peterson Dairy, like most other dairies of the time, made milk deliveries to homes. This vehicle is a 1920s Dodge Delivery. (Diane Peterson Graham and Maynard "Pete" H. Peterson.)

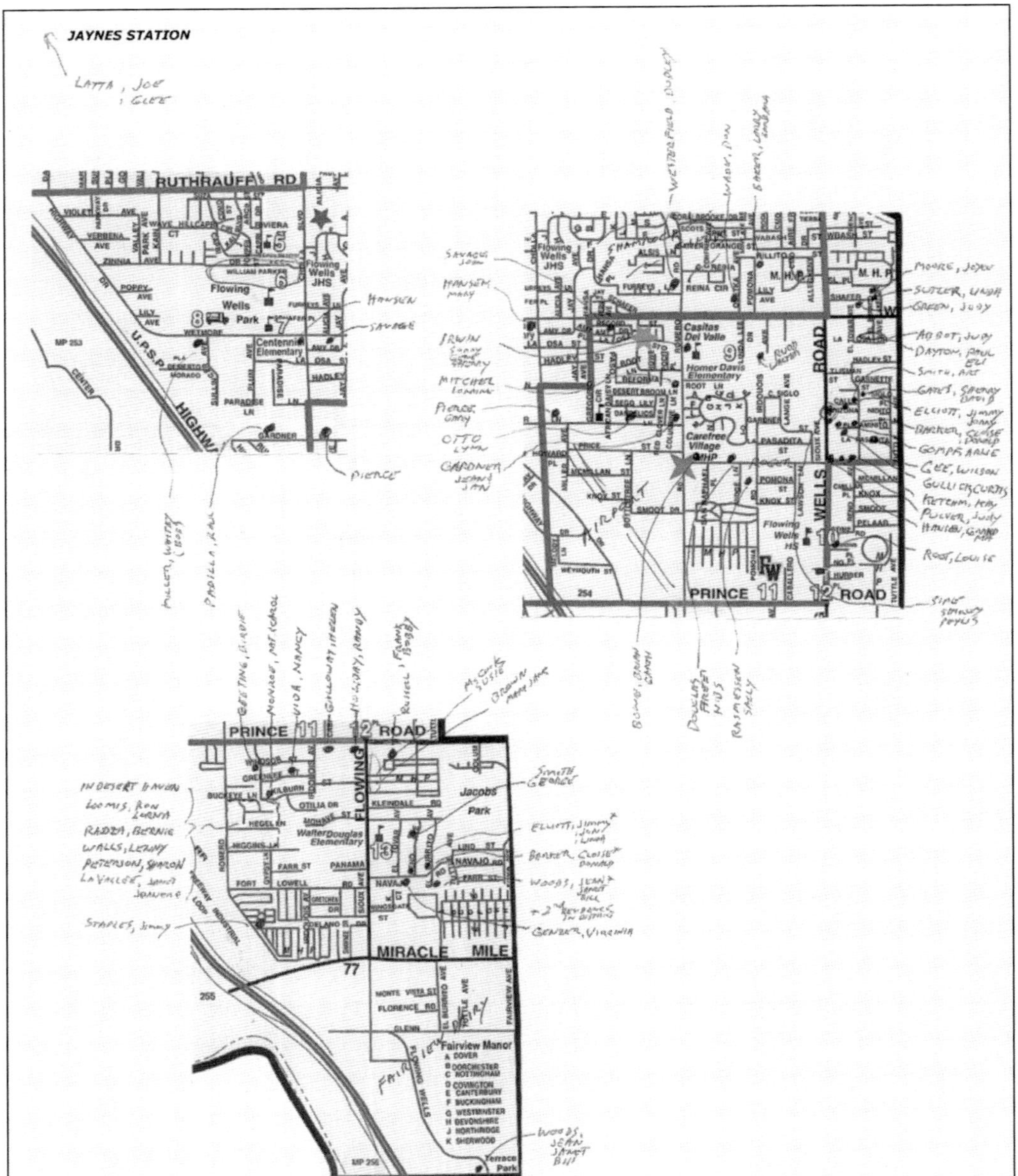

FWHS's first full-size graduating class was in 1957. This detailed listing of the families and their homes is overlaid on a current-day map of the Flowing Wells area. (James Elliott.)

Three

Laguna School History

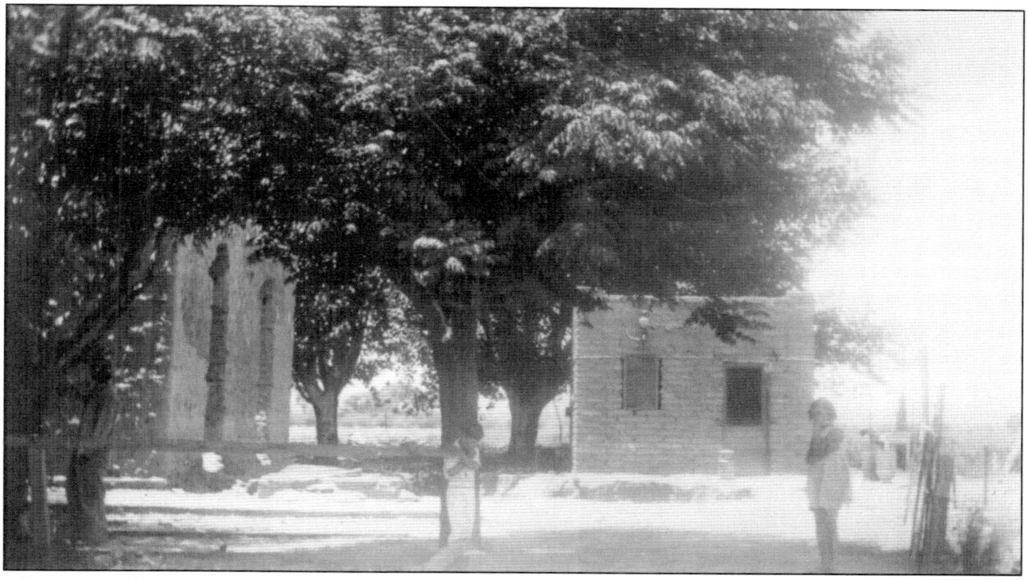

The original Laguna School was built with adobe bricks in 1889. The building is now used as a residence just north of the present-day Laguna Elementary School. It is pictured here around 1930. (FWUSD.)

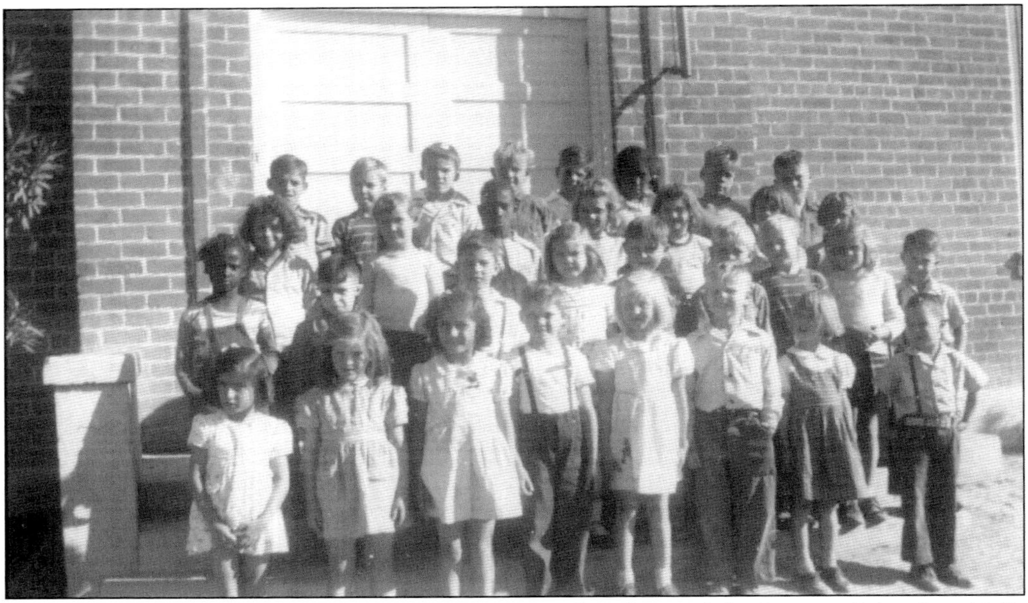

First-, second-, and third-grade students are pictured in front of the Laguna School in 1948. The redbrick Laguna School built in 1915 and was torn down in 1975. A modern building was constructed in 1963 and is part of the present Laguna Elementary School. (FWUSD.)

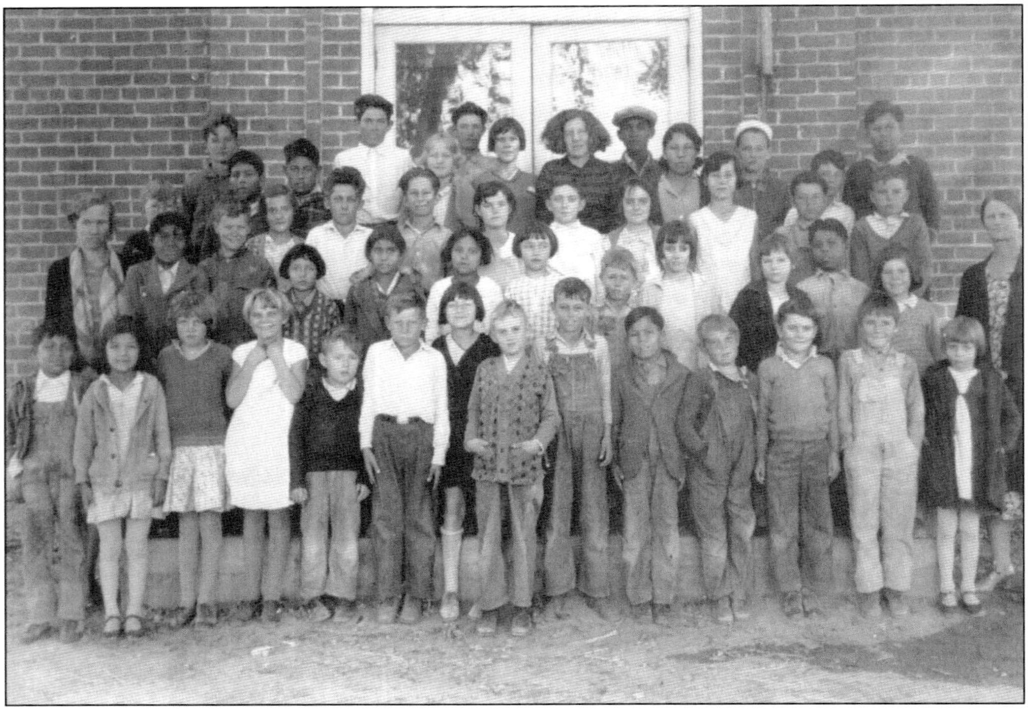

This Laguna School class picture was taken in 1930. Corinne Sptizer, seventh from left in the first row, donated this photograph to the present-day Laguna Elementary School. (FWUSD.)

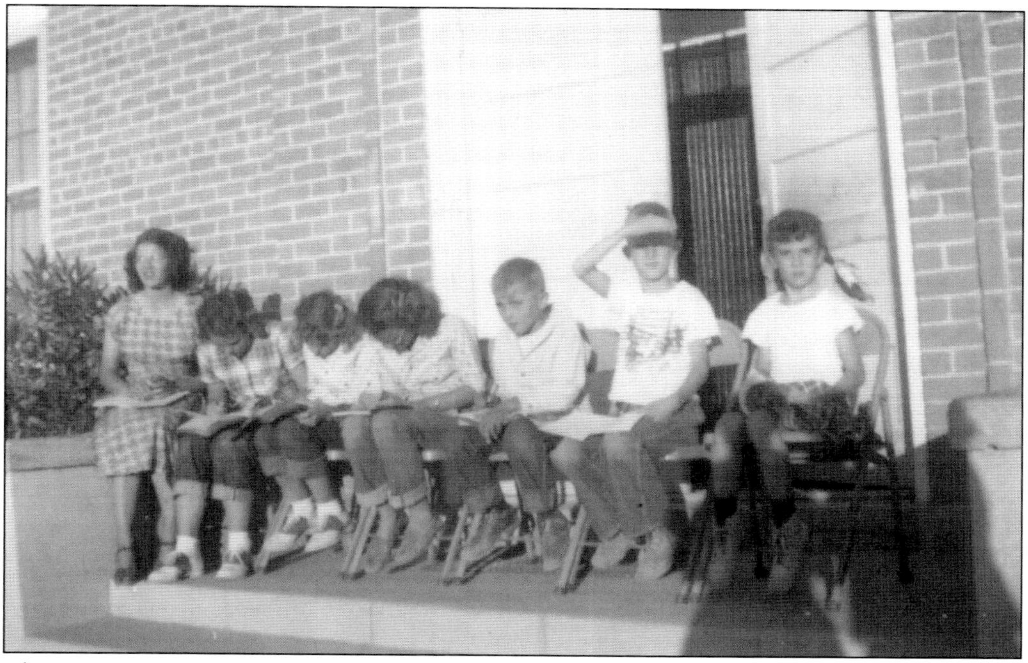

These Laguna School students studying outdoors in 1950 are, from left to right, Betty Ertle, JoAnne Bennett, Willene Bennett, Ruth Shumaker, Howard King, Joe Latta, and Glee Latta. (FWUSD.)

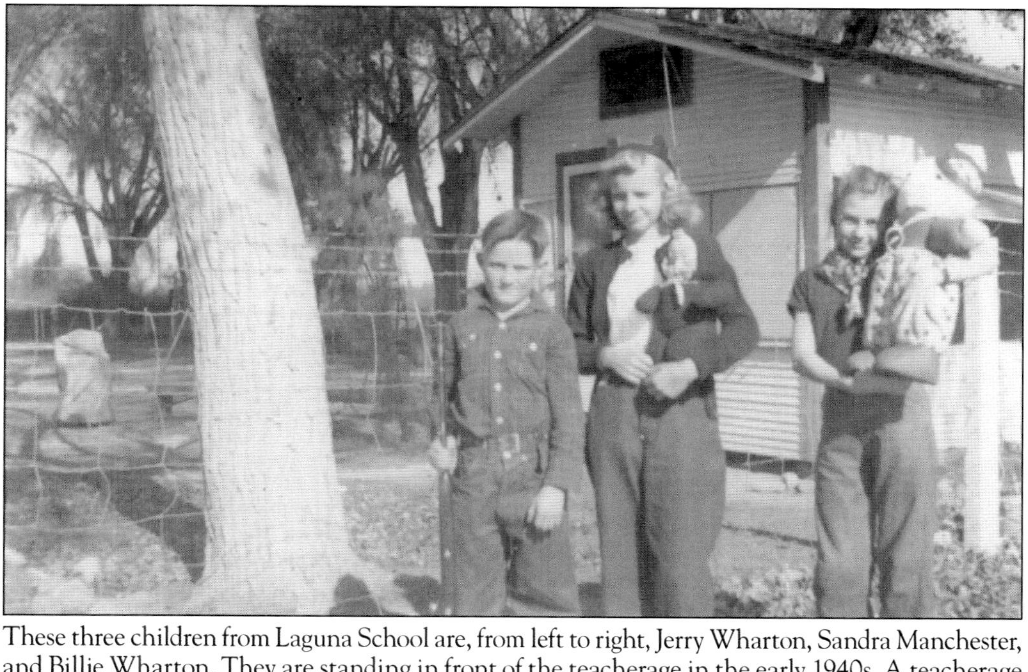

These three children from Laguna School are, from left to right, Jerry Wharton, Sandra Manchester, and Billie Wharton. They are standing in front of the teacherage in the early 1940s. A teacherage was a house or accommodation provided for a teacher by the school. (FWUSD.)

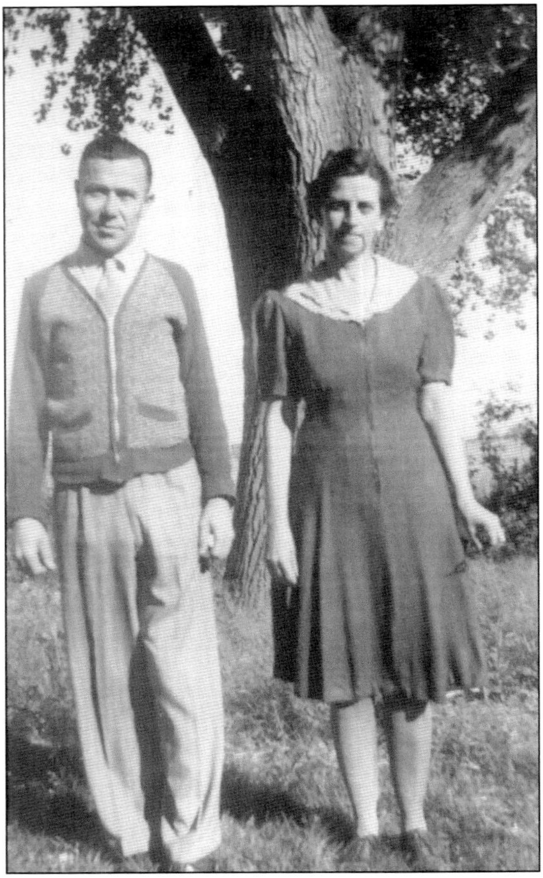

The two-room school was taught by these two Laguna teachers pictured in the early 1940s. First through third grades on the north end were taught by Virginia Burton, and fourth through eighth grades on the south end were taught by William Wharton. (FWUSD.)

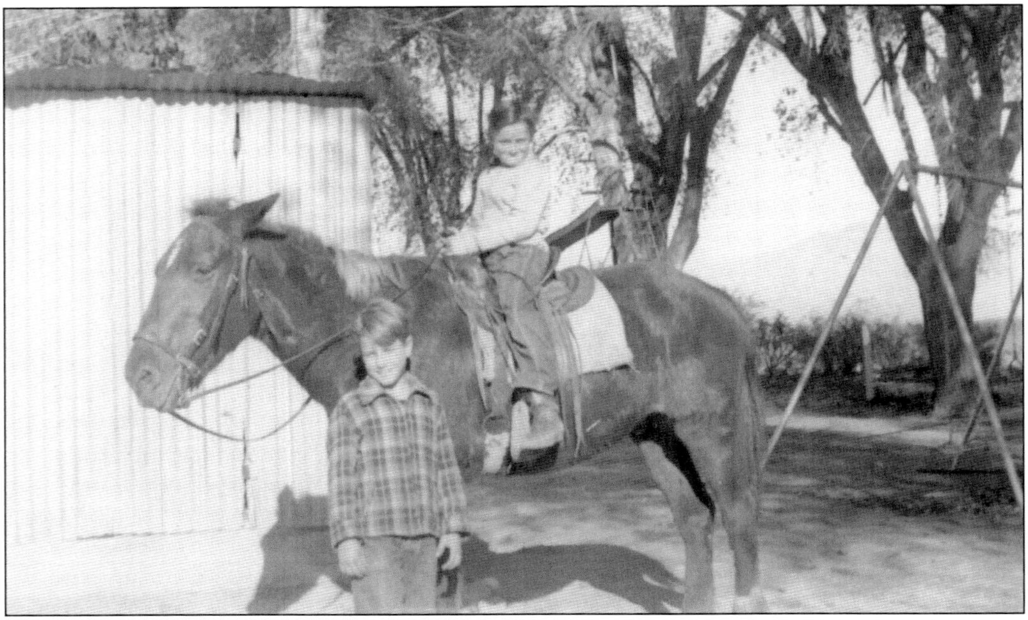

Winston Shumaker and Jerry Wharton (on the horse) are pictured in front of the pumphouse on the Laguna playground in the early 1940s. (FWUSD.)

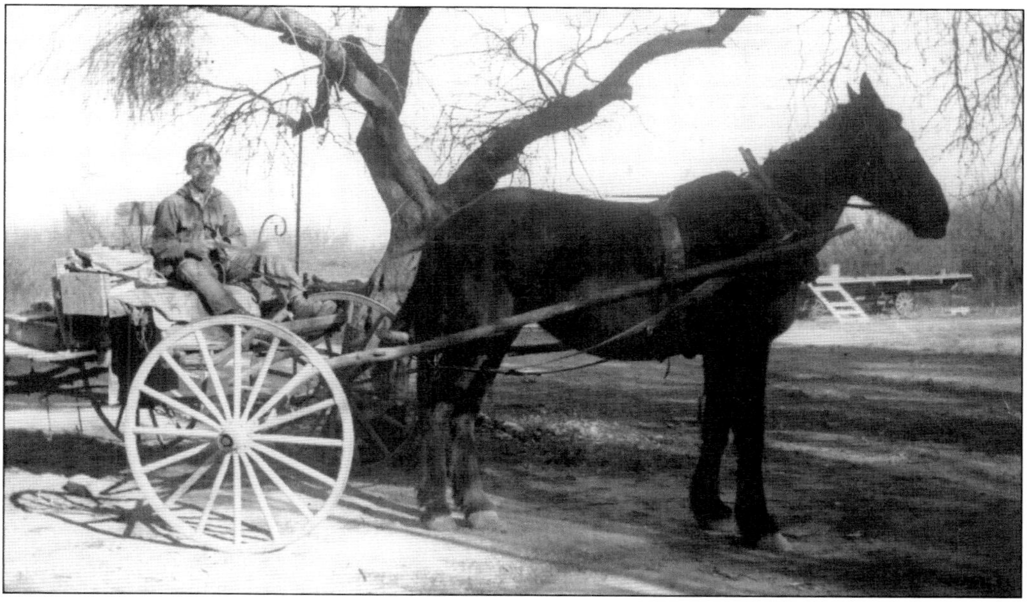

Children in the 1940s rode horse-drawn buckboards to get to school. This boy attended the Laguna School. Harold "Bob" Butler lived on the old Harper farm in 1938 where the Lu Lu Walker School (amphitheater) is today on Roller Coaster Road, and remembers in second grade having to turn back when the Rillito River was running. (FWUSD.)

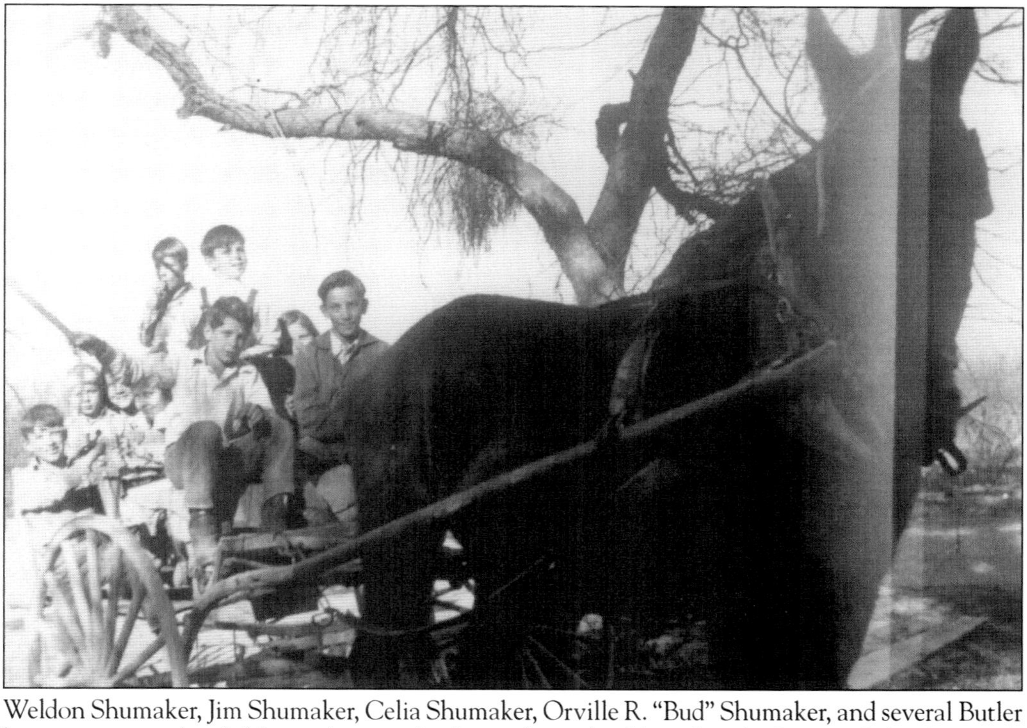

Weldon Shumaker, Jim Shumaker, Celia Shumaker, Orville R. "Bud" Shumaker, and several Butler boys are among the children riding on this horse-drawn buckboard to the Laguna School in 1943. Shumaker Drive, located just north of the Rillito River off Camino De La Tierra, is believed to be named after the Shumaker family. (FWUSD.)

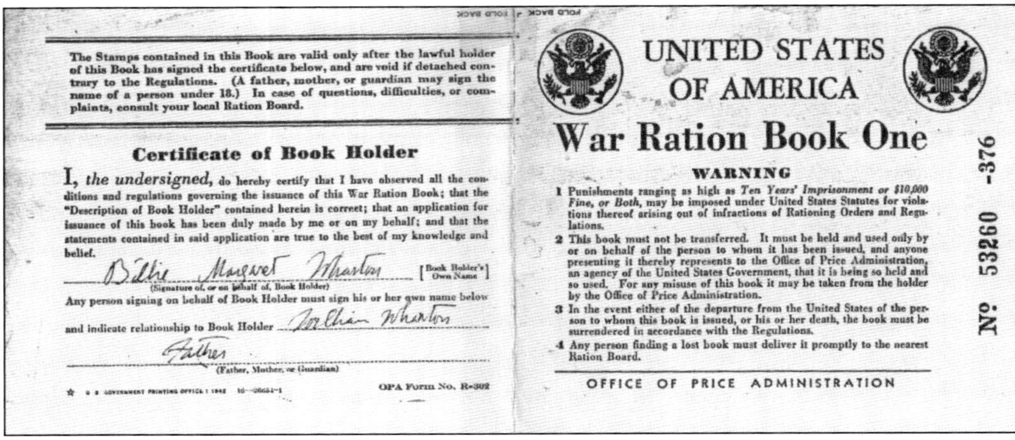

The Stamps contained in this Book are valid only after the lawful holder of this Book has signed the certificate below, and are void if detached contrary to the Regulations. (A father, mother, or guardian may sign the name of a person under 18.) In case of questions, difficulties, or complaints, consult your local Ration Board.

Certificate of Book Holder

I, *the undersigned,* do hereby certify that I have observed all the conditions and regulations governing the issuance of this War Ration Book; that the "Description of Book Holder" contained herein is correct; that an application for issuance of this book has been duly made by me or on my behalf; and that the statements contained in said application are true to the best of my knowledge and belief.

Billie Margaret Wharton [Book Holder's Own Name]
(Signature of, or on behalf of, Book Holder)

Any person signing on behalf of Book Holder must sign his or her own name below and indicate relationship to Book Holder William Wharton

Father
(Father, Mother, or Guardian)

OPA Form No. R-302

UNITED STATES OF AMERICA

War Ration Book One

Nº 53260 -376

WARNING

1 Punishments ranging as high as *Ten Years' Imprisonment or $10,000 Fine, or Both,* may be imposed under United States Statutes for violations thereof arising out of infractions of Rationing Orders and Regulations.

2 This book must not be transferred. It must be held and used only by or on behalf of the person to whom it has been issued, and anyone presenting it thereby represents to the Office of Price Administration, an agency of the United States Government, that it is being so held and so used. For any misuse of this book it may be taken from the holder by the Office of Price Administration.

3 In the event either of the departure from the United States of the person to whom this book is issued, or his or her death, the book must be surrendered in accordance with the Regulations.

4 Any person finding a lost book must deliver it promptly to the nearest Ration Board.

OFFICE OF PRICE ADMINISTRATION

Laguna teacher Virginia Burton organized the war ration books for the Jaynes Station area. This ration book was issued to Billie Margaret Wharton in 1942 when she was 13 years old. (FWUSD.)

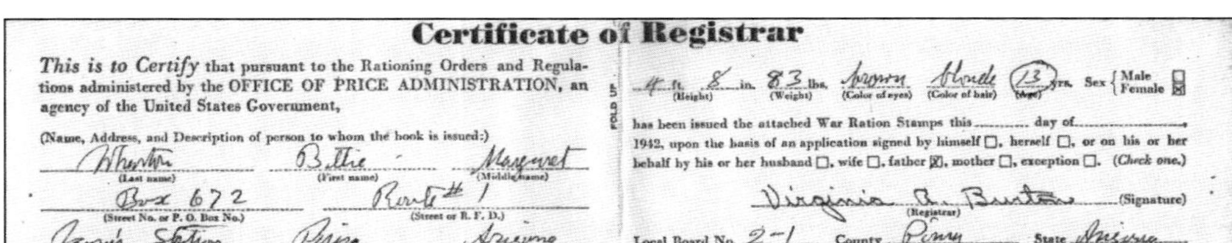

Certificate of Registrar

This is to Certify that pursuant to the Rationing Orders and Regulations administered by the OFFICE OF PRICE ADMINISTRATION, an agency of the United States Government,

(Name, Address, and Description of person to whom the book is issued:)

Wharton (Last name) Billie (First name) Margaret (Middle name)

Box 672 (Street No. or P. O. Box No.) Route #1 (Street or R. F. D.)

Jaynes Station (City or town) Pima (County) Arizona (State)

4 ft. 8 in. (Height) 83 lbs. (Weight) brown (Color of eyes) blonde (Color of hair) 13 yrs. (Age) Sex {Male / Female ☒}

has been issued the attached War Ration Stamps this .... day of ...., 1942, upon the basis of an application signed by himself ☐, herself ☐, or on his or her behalf by his or her husband ☐, wife ☐, father ☒, mother ☐, exception ☐. (*Check one.*)

Virginia B. Burton (Registrar) (Signature)

Local Board No. 2-1 County Pima State Arizona

Stamps must not be detached except in the presence of the retailer, his employee, or person authorized by him to make delivery.

War ration books were issued during World War II. This is book one of a series issued during the war. The book contained removable stamps for rationed items like sugar, meat, cooking oil, and canned goods. People submitted these stamps at a grocer. (FWUSD.)

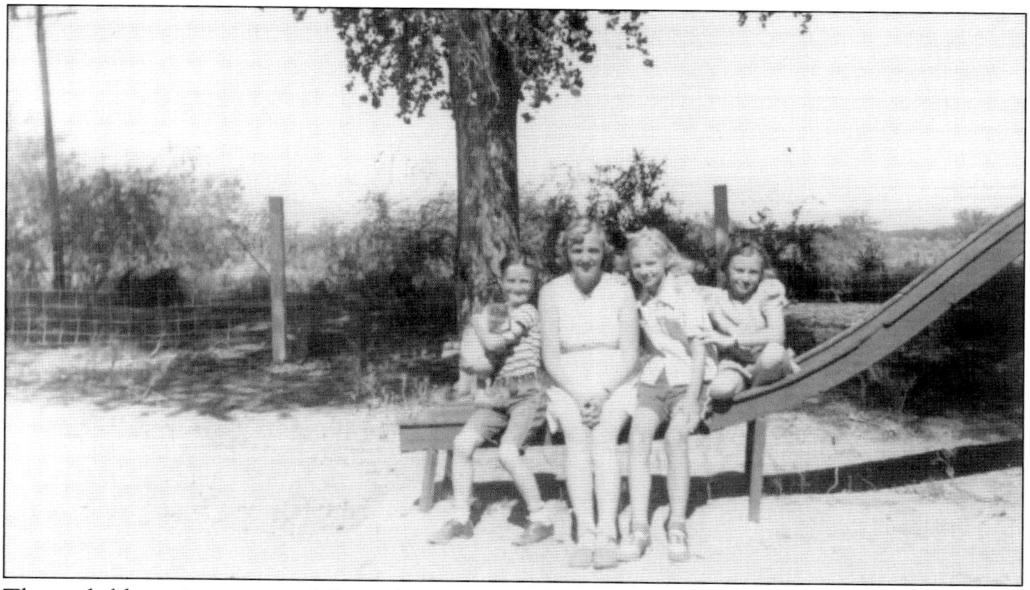

These children sitting on a slide at the Laguna School playground in the early 1940s are, from left to right, Jerry Wharton, Grace Manchester, Sandra Manchester, and Billie Wharton. Myreel Smith Lewis (not pictured) remembers "eating our sack lunches out under the big trees. Our sandwiches were wrapped in wax paper so after lunch we'd take our wax paper and sit on it on the slide. Of course, the heat on the metal would melt the wax, so our ride on the slide was lickety-split. Fun!" (FWUSD.)

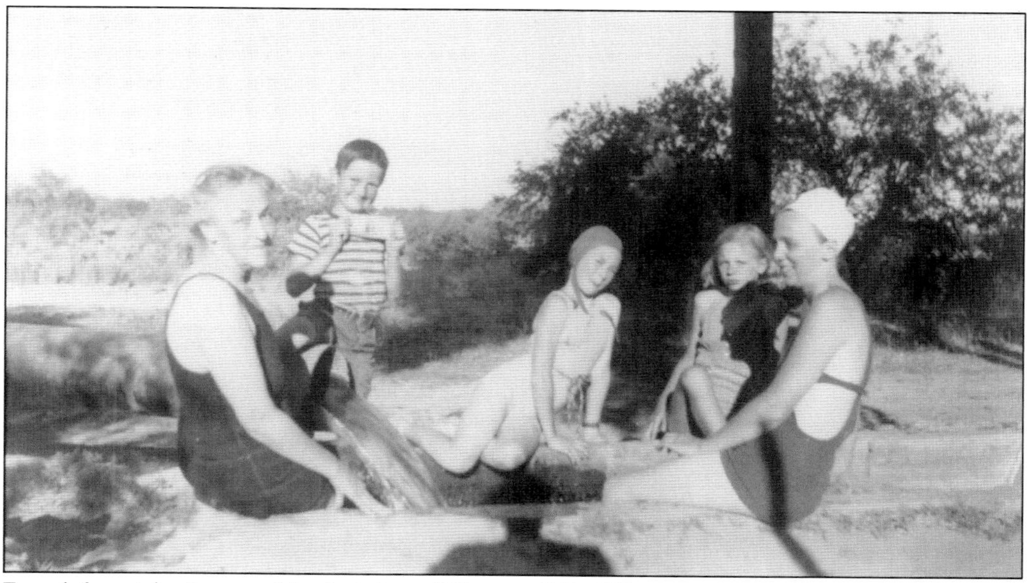

From left to right, Jeanette Murdock, Jerry Wharton, Billie Wharton, Grace Manchester, and Sandra Manchester cool off at an irrigation pump near Laguna School in the early 1940s. (FWUSD.)

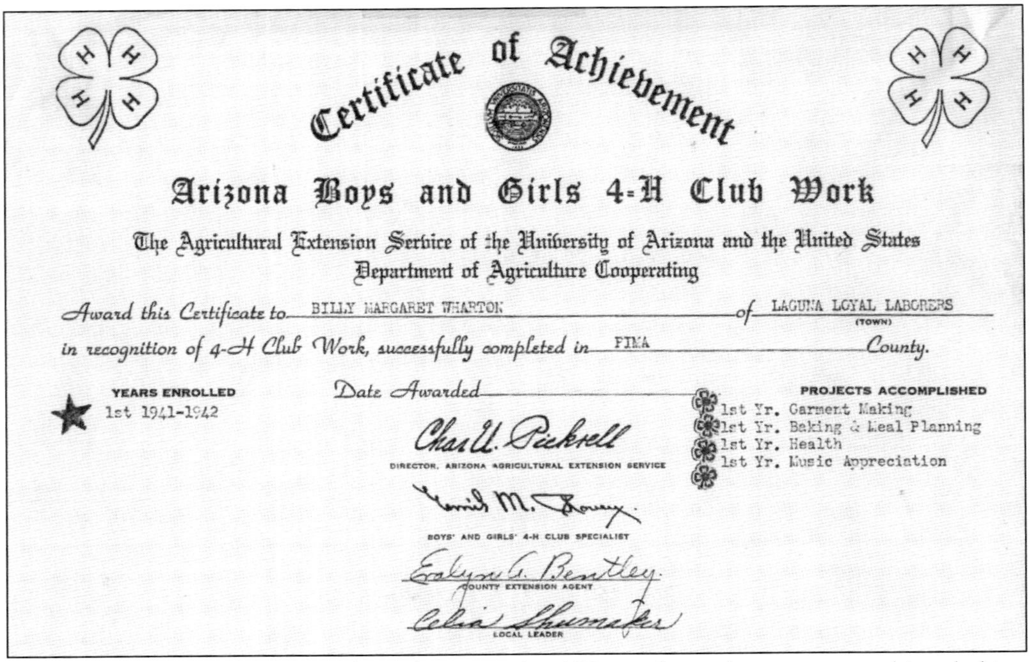

Certificate of Achievement

Arizona Boys and Girls 4-H Club Work

The Agricultural Extension Service of the University of Arizona and the United States Department of Agriculture Cooperating

Award this Certificate to BILLY MARGARET WHARTON of LAGUNA LOYAL LABORERS (TOWN)

in recognition of 4-H Club Work, successfully completed in PIMA County.

YEARS ENROLLED
1st 1941-1942

Date Awarded

Chas U. Pickrell
DIRECTOR, ARIZONA AGRICULTURAL EXTENSION SERVICE

BOYS' AND GIRLS' 4-H CLUB SPECIALIST

Evalyn A. Bentley
COUNTY EXTENSION AGENT

Celia Shumaker
LOCAL LEADER

PROJECTS ACCOMPLISHED
1st Yr. Garment Making
1st Yr. Baking & Meal Planning
1st Yr. Health
1st Yr. Music Appreciation

In 1941–1942, Billie Margaret Shumaker earned a 4-H certificate for garment making, baking and meal planning, health, and music appreciation. Celia Shumaker was selected in 1942 as the best all-around 4-H Club girl in Arizona and was selected to attend the annual National 4-H Club Conference in Chicago. When she returned to Tucson, she gave a presentation to the kids in the community about her adventure. (FWUSD.)

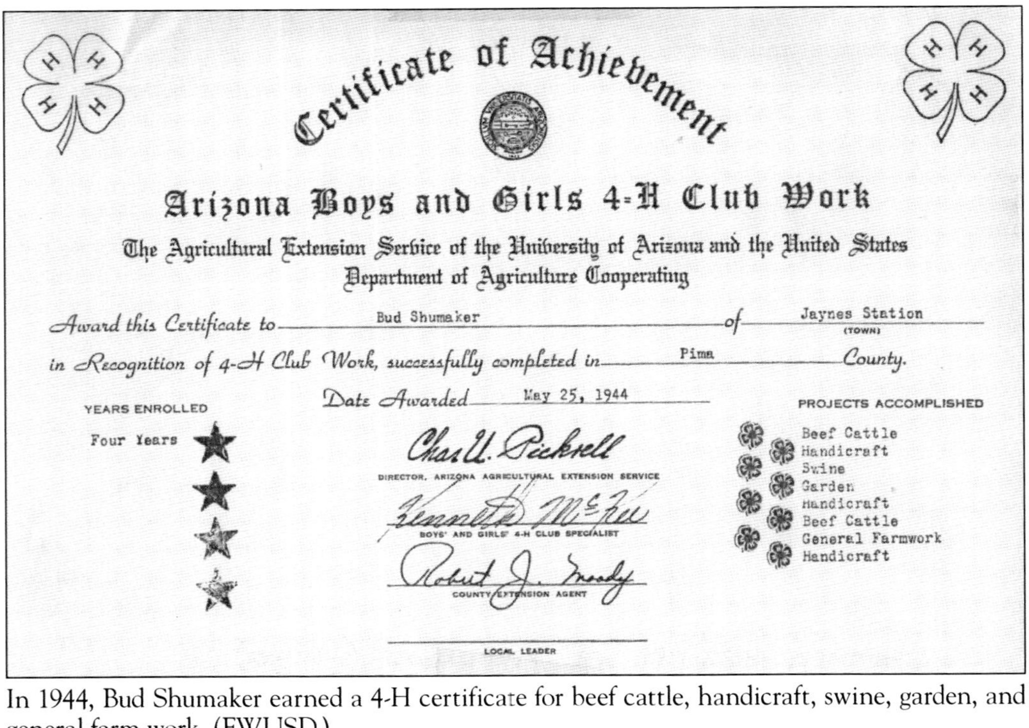

Certificate of Achievement

Arizona Boys and Girls 4-H Club Work

The Agricultural Extension Service of the University of Arizona and the United States Department of Agriculture Cooperating

Award this Certificate to Bud Shumaker of Jaynes Station (TOWN)

in Recognition of 4-H Club Work, successfully completed in Pima County.

YEARS ENROLLED
Four Years

Date Awarded May 25, 1944

Chas U. Pickrell
DIRECTOR, ARIZONA AGRICULTURAL EXTENSION SERVICE

Kenneth McKee
BOYS' AND GIRLS' 4-H CLUB SPECIALIST

Robert J. Moody
COUNTY EXTENSION AGENT

LOCAL LEADER

PROJECTS ACCOMPLISHED
Beef Cattle
Handicraft
Swine
Garden
Handicraft
Beef Cattle
General Farmwork
Handicraft

In 1944, Bud Shumaker earned a 4-H certificate for beef cattle, handicraft, swine, garden, and general farm work. (FWUSD.)

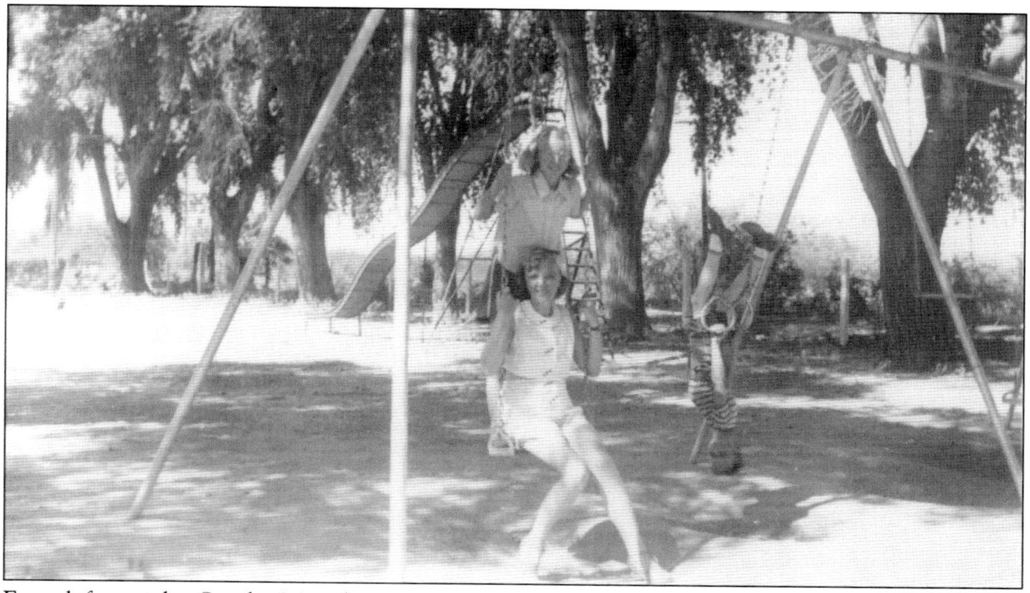

From left to right, Sandra Manchester, Grace Manchester, and Jerry Wharton are on the swings at the Laguna School in the early 1940s. (FWUSD.)

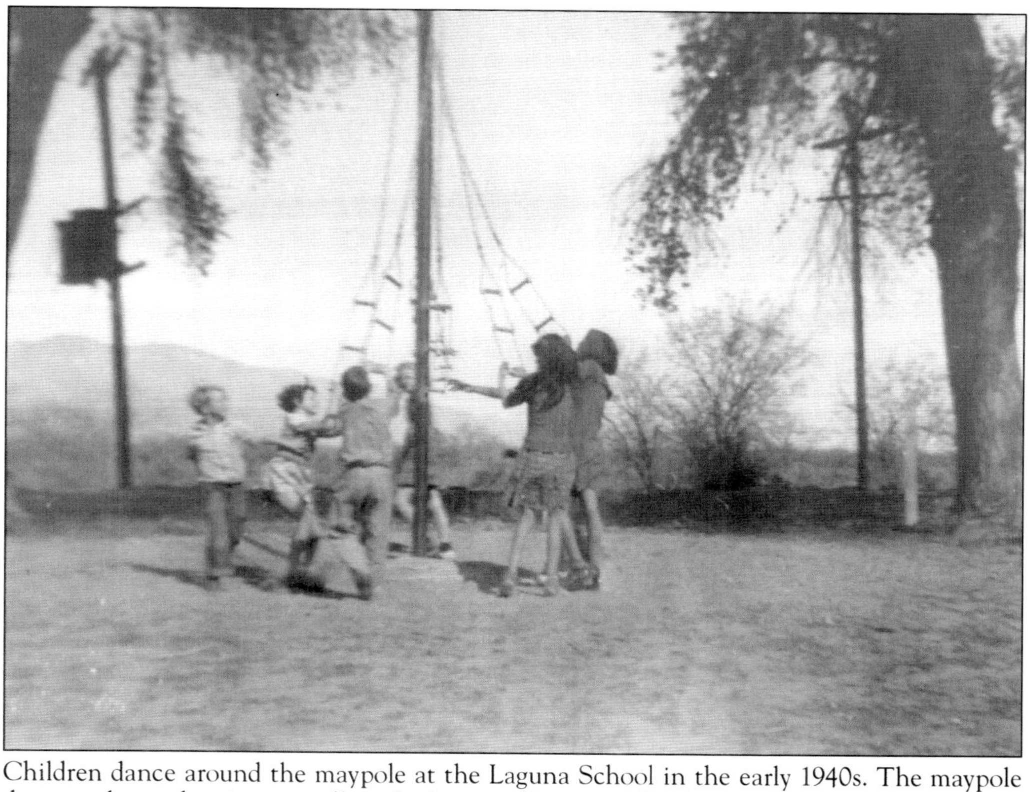

Children dance around the maypole at the Laguna School in the early 1940s. The maypole dance welcomed spring as well as the hopes of bringing fertility to the land, livestock, and people. (FWUSD.)

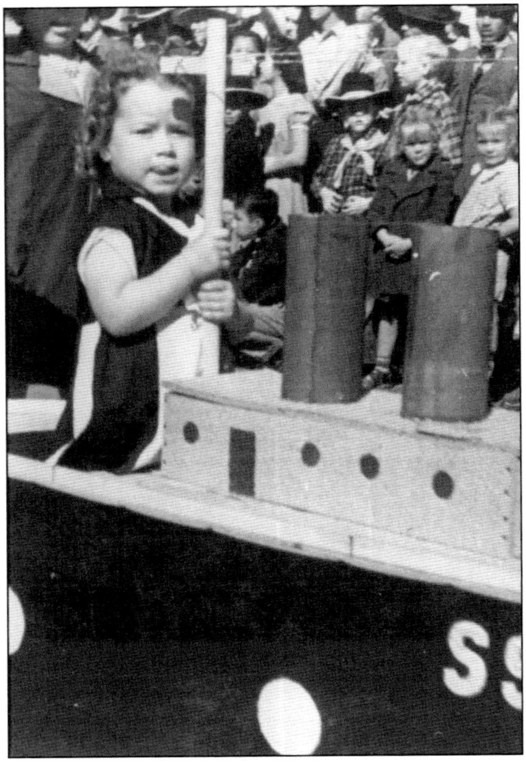

Three-year-old Ruth Shumaker rides on the SS Laguna float in the 1941 Rodeo Day Parade. The parade for the Tucson Rodeo started in 1925 and celebrated its 93rd year in February 2018. Ruth's parents James A. and Susan Edna Bond Shumaker farmed in the Jaynes Station area, where James was active on the Laguna School Board. Their son Winston Shumaker was a member of the Laguna board when it was dissolved after annexation in 1964. (FWUSD.)

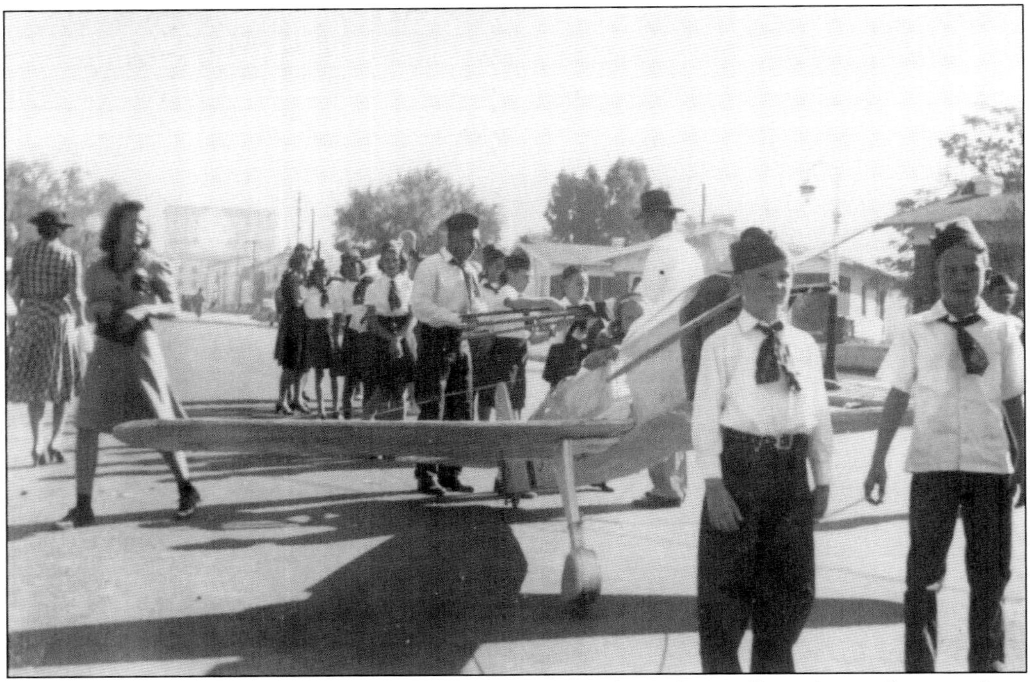

Laguna School students Jerry Wharton and Winston Shumaker walk in front of an airplane in the Rodeo Day Parade in 1942. The Gilpin Airport was in operation and World War II was occurring at this time. Gary Abrams (FWHS 1970) believes the plane may have been fashioned after a PT-22 pilot trainer plane. (FWUSD.)

The graduation ceremony for eighth-grade students in the 20 area rural schools of Pima County was held at Wakefield Junior High School on May 22, 1942. The program's front cover is pictured. At the time, Flowing Wells and Laguna were separate school districts. (FWUSD.)

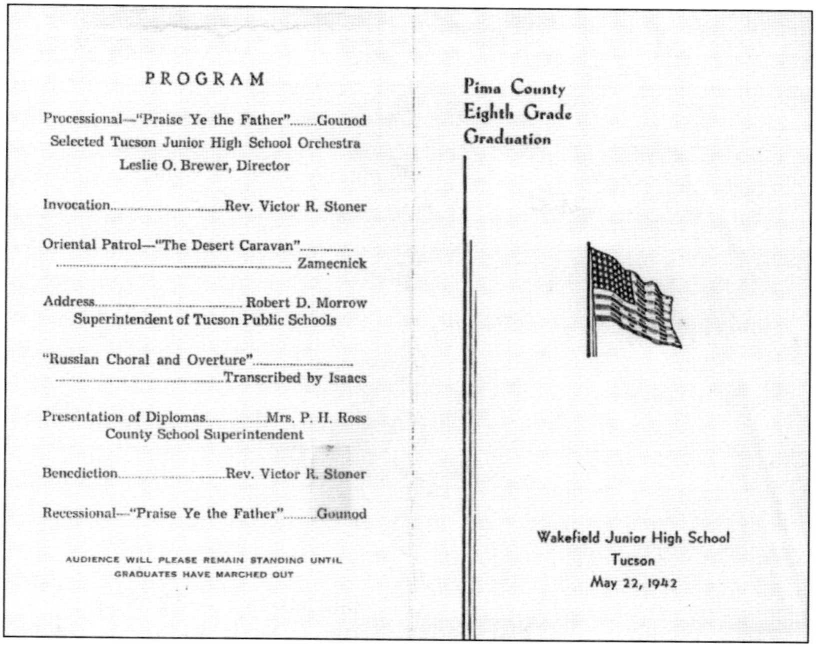

PROGRAM

Processional—"Praise Ye the Father"........Gounod
Selected Tucson Junior High School Orchestra
Leslie O. Brewer, Director

Invocation..............................Rev. Victor R. Stoner

Oriental Patrol—"The Desert Caravan"..............
....Zamecnick

Address..............................Robert D. Morrow
Superintendent of Tucson Public Schools

"Russian Choral and Overture"..........................
....Transcribed by Isaacs

Presentation of Diplomas.................Mrs. P. H. Ross
County School Superintendent

Benediction..............................Rev. Victor R. Stoner

Recessional—"Praise Ye the Father"..........Gounod

AUDIENCE WILL PLEASE REMAIN STANDING UNTIL GRADUATES HAVE MARCHED OUT

Pima County
Eighth Grade
Graduation

Wakefield Junior High School
Tucson
May 22, 1942

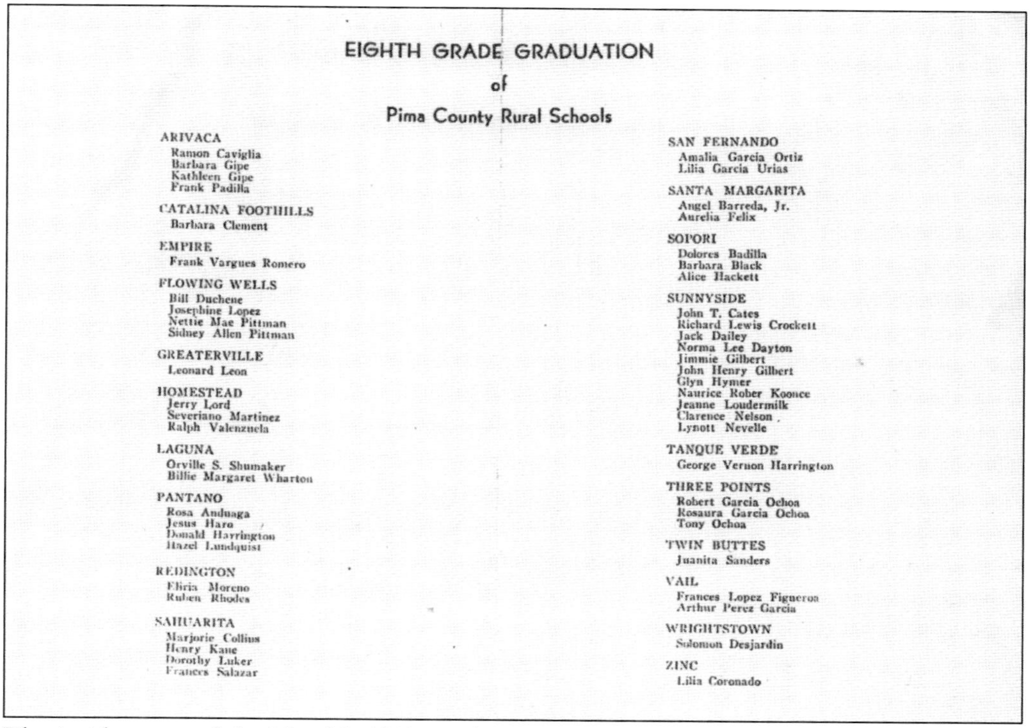

EIGHTH GRADE GRADUATION
of
Pima County Rural Schools

ARIVACA
Ramon Caviglia
Barbara Gipe
Kathleen Gipe
Frank Padilla

CATALINA FOOTHILLS
Barbara Clement

EMPIRE
Frank Vargues Romero

FLOWING WELLS
Bill Duchene
Josephine Lopez
Nettie Mae Pittman
Sidney Allen Pittman

GREATERVILLE
Leonard Leon

HOMESTEAD
Jerry Lord
Severiano Martinez
Ralph Valenzuela

LAGUNA
Orville S. Shumaker
Billie Margaret Wharton

PANTANO
Rosa Anduaga
Jesus Haro
Donald Harrington
Hazel Lundquist

REDINGTON
Eliria Moreno
Ruben Rhodes

SAHUARITA
Marjorie Collins
Henry Kane
Dorothy Luker
Frances Salazar

SAN FERNANDO
Amalia Garcia Ortiz
Lilia Garcia Urias

SANTA MARGARITA
Angel Barreda, Jr.
Aurelia Felix

SOPORI
Dolores Badilla
Barbara Black
Alice Hackett

SUNNYSIDE
John T. Cates
Richard Lewis Crockett
Jack Dailey
Norma Lee Dayton
Jimmie Gilbert
John Henry Gilbert
Glyn Hymer
Naurice Rober Koonce
Jeanne Loudermilk
Clarence Nelson
Lynott Nevelle

TANQUE VERDE
George Vernon Harrington

THREE POINTS
Robert Garcia Ochoa
Rosaura Garcia Ochoa
Tony Ochoa

TWIN BUTTES
Juanita Sanders

VAIL
Frances Lopez Figueroa
Arthur Perez Garcia

WRIGHTSTOWN
Solomon Desjardin

ZINC
Lilia Coronado

The inside pages of the May 22, 1942, graduation program are pictured here. Laguna School had two students: Orville Shumaker and Billie Margaret Wharton. Flowing Wells School had four students: Bill Duchene, Josephine Lopez, Nettie Mea Pittman, and Sidney Allen Pittman. Laguna School District was annexed into the Flowing Wells District at the request of Laguna residents in 1964. (FWUSD.)

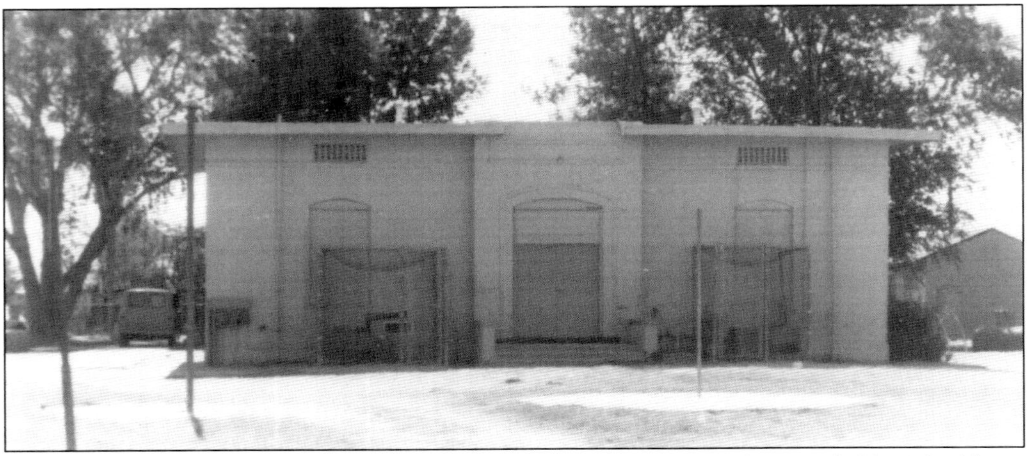

The second Laguna School, pictured in the 1960s, was built in 1915. On March 23, 1961, fifteen electors signed a petition for the Laguna School District to be annexed into the Flowing Wells School District. The Flowing Wells School Board accepted the petition on the same day, and the petition was filed on March 24. On May 2, 1961, the vote to be annexed failed 106-10 due to the ballot asking if Laguna citizens wanted to assume a proportionate share of liability for outstanding bond debt of the Flowing Wells School District. On Saturday, February 29, 1964, the Laguna School District requested to be annexed for a second time. On March 24, 1964, with a vote of 161-116, citizens of the Laguna School District No. 17, voted to be annexed into the Flowing Wells School District No. 8. This vote more than doubled the Flowing Wells district. Before being annexed, students chose between Tucson, Amphitheater, or Flowing Wells High Schools to attend ninth through twelfth grades. (FWUSD.)

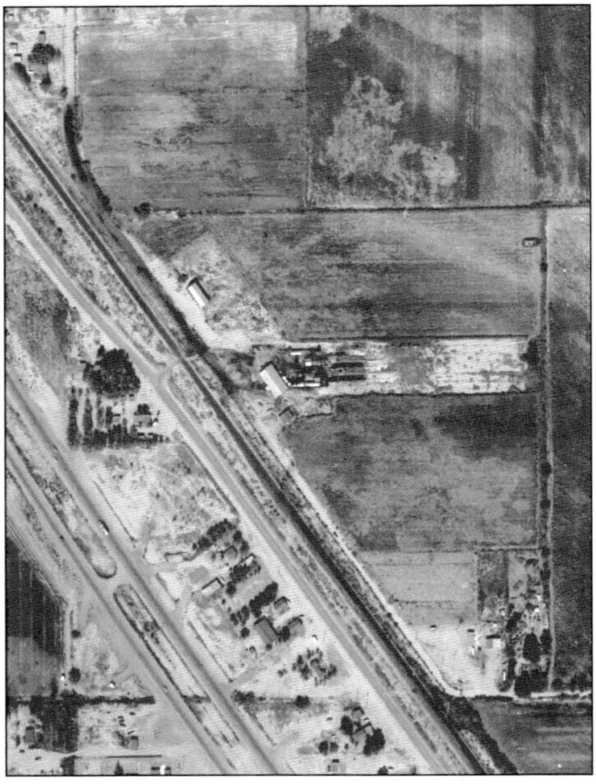

Jaynes Station, pictured in 1953, was between the railroad tracks and Highway 84. Don McMurrich remembers his father driving cattle to load on the train at Jaynes Station in the 1950s. Laguna School is at lower right at the intersection of today's Diamond Street and Shannon Road. Shannon Road is named after Lester Shannon, who purchased land on the south bank of the Rillito and operated a dairy and farmed hay, alfalfa, and cotton from 1937 to 1957. He also worked for what is now the Pima County Department of Transportation from 1945 to 1958. Curtis Road is named after farmer Ammon Ralph Curtis, who purchased land from Lester Shannon. (Cooper Aerial Survey and authors.)

Four

GILPIN FIELD

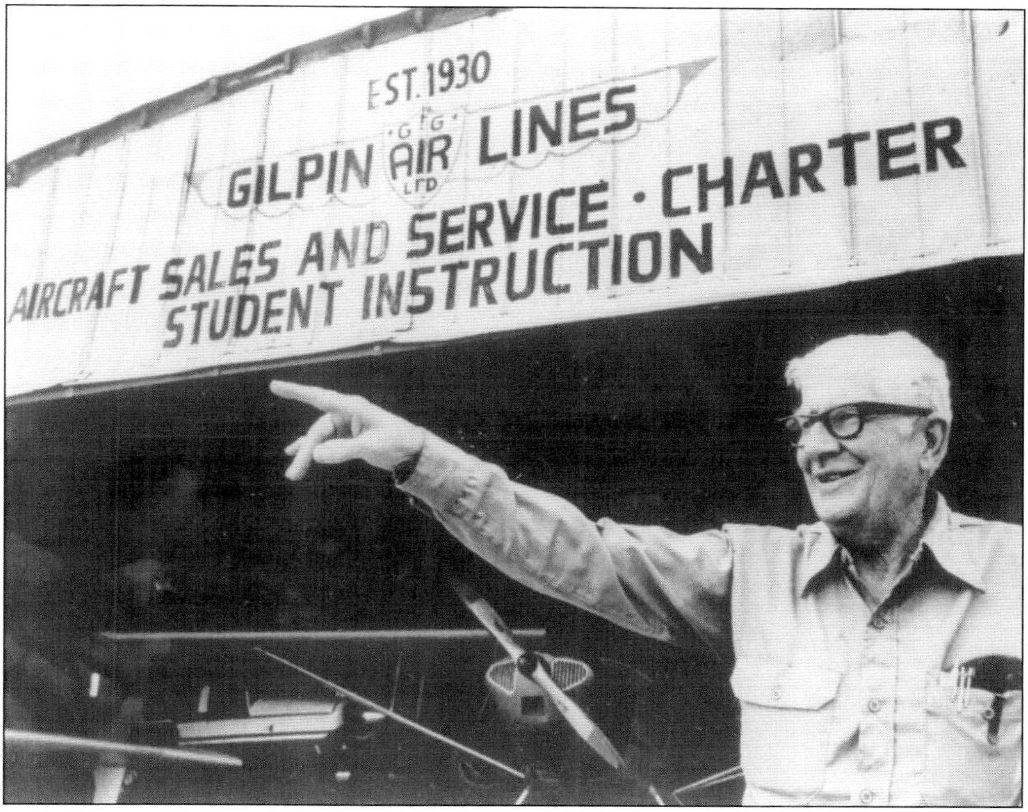

Walter Douglas Jr. is pictured in front of the Gilpin Airlines/Freeway Airport hangar in 1975. This hangar is still standing on Romero Road near Prince Road. In 1956, Douglas sold 224 acres of the airport to Gail Hummel Associates (negotiated by Heinz Wegner and Associates). The airport closed in February 1957 and was reopened by Wegner in July 1957 operating on a limited basis. Wegner, a former test pilot for Lockheed Aircraft Company, worked for Douglas in the 1930s. In 1958, Arthur N. Pack, owner of the Ghost Ranch Lodge at 801 West Casa Grande Highway (now called Miracle Mile) purchased 100 of the Gilpin acres, not including any buildings or hangars. New hangars were built on Prince Road and were reopened as Freeway Airport. G&G Airlines stands for Gilpin and Greenway. (Jeretta Douglas.)

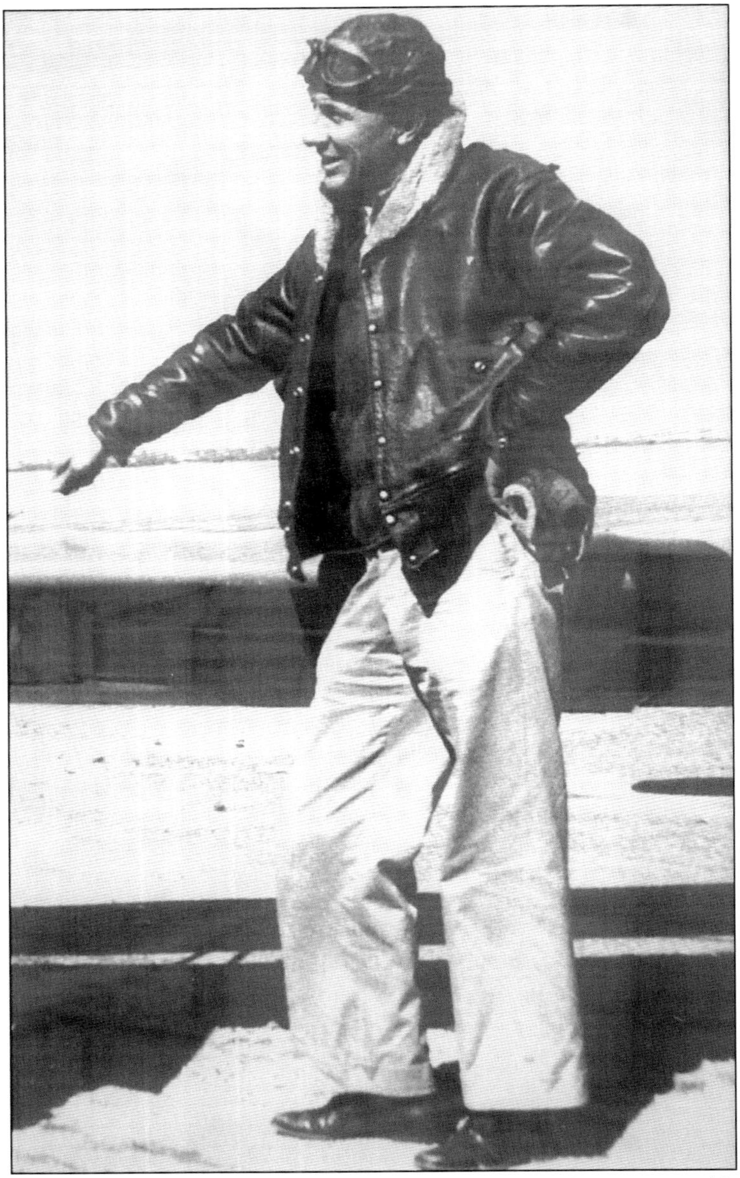

Walter Douglas Jr. is pictured around 1945 in his signature flight jacket, His grandfather Dr. James Douglas was a mining engineer and metallurgist commissioned by the Phelps Dodge Corporation in 1881. He served as manager of the Copper Queen Consolidated Mining Company from 1885 to 1917 and president of Phelps Dodge from 1909 to 1918. The town of Douglas is named after him. Walter Douglas Sr. was prominent in railroad development in southern Arizona and Mexico. He served as president and chairman of Phelps Dodge and retired in 1930. Douglas Jr. was not interested in mining like his father and grandfather. In 1928, he started flying and established Gilpin Airlines in San Francisco. His godmother, Isabella Greenway, encouraged Douglas to purchase 400 acres of farmland near Roger Road and plant cotton. Greenway was Arizona's first US congresswoman. She founded the Arizona Inn and served as a bridesmaid at the wedding of Eleanor and Franklin D. Roosevelt. Douglas's love of aviation led him in 1938 to clear the cottonfields and start building G&G Airlines. He offered charter services and flying instruction. He sold the airport land in 1956 due to urban development. (Jeretta Douglas.)

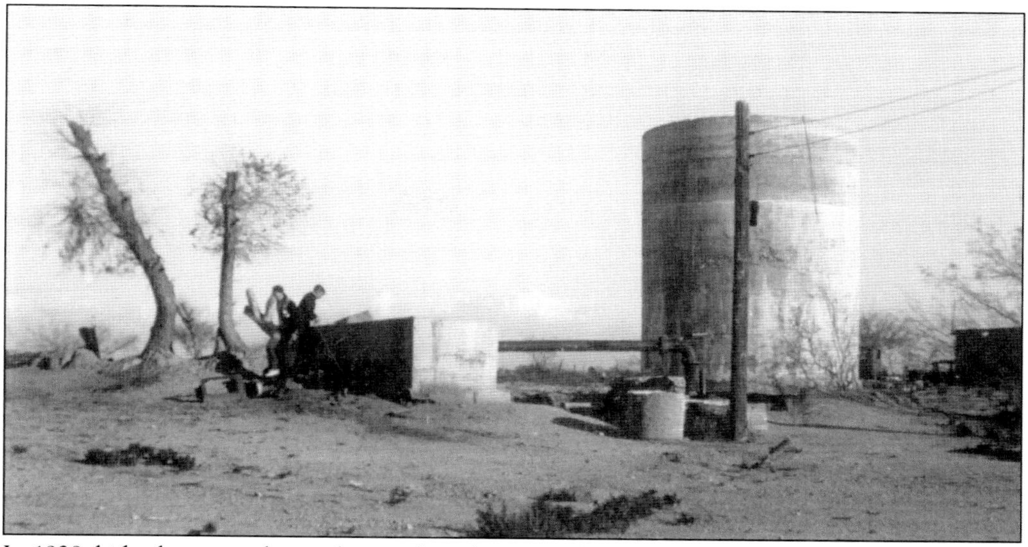

In 1938, kids play around a cattle trough and water well tank near Prince and Romero Roads on the land where Gilpin Airport was constructed. (Arizona Historical Society.)

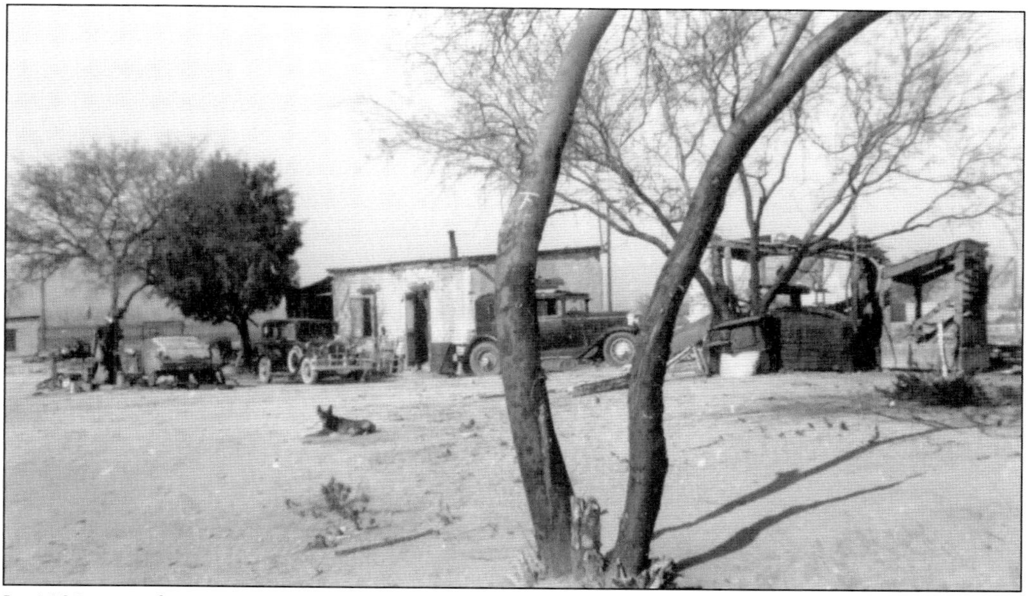

In 1938, very few structures were located on the property at Prince and Romero Roads where Walter Douglas Jr. built G&G Airlines. Construction on the airport started in 1938 and was completed in 1941. (Arizona Historical Society.)

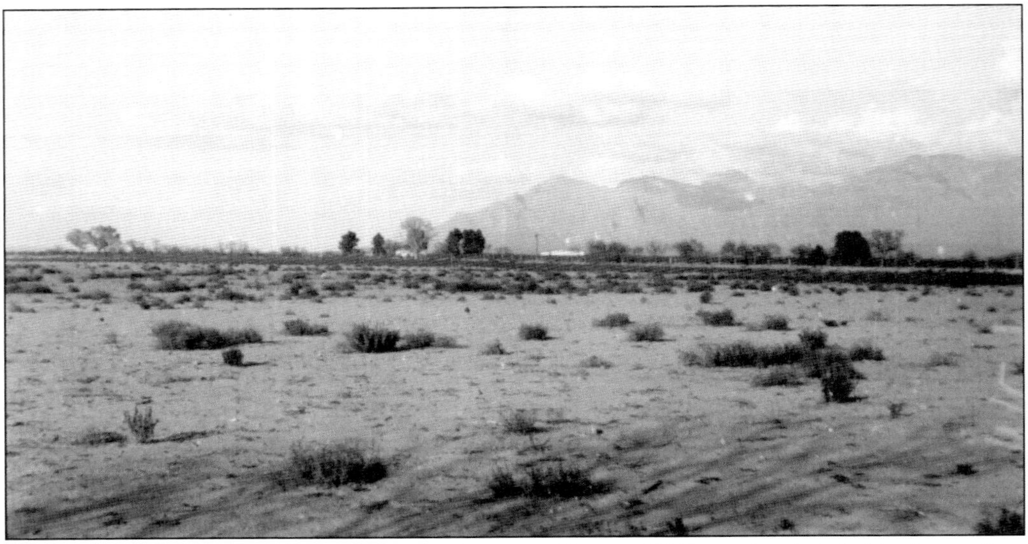

Around 1938, prior to World War II, the area known as Flowing Wells was barely inhabited, providing an unobstructed view of the Catalina Mountains to the north, as seen around 1938. Around 1940, the irrigation district's wells were drilled, Gilpin Airport grew, roads were built, residential areas were developed, and shopping centers and schools followed. (Arizona Historical Society.)

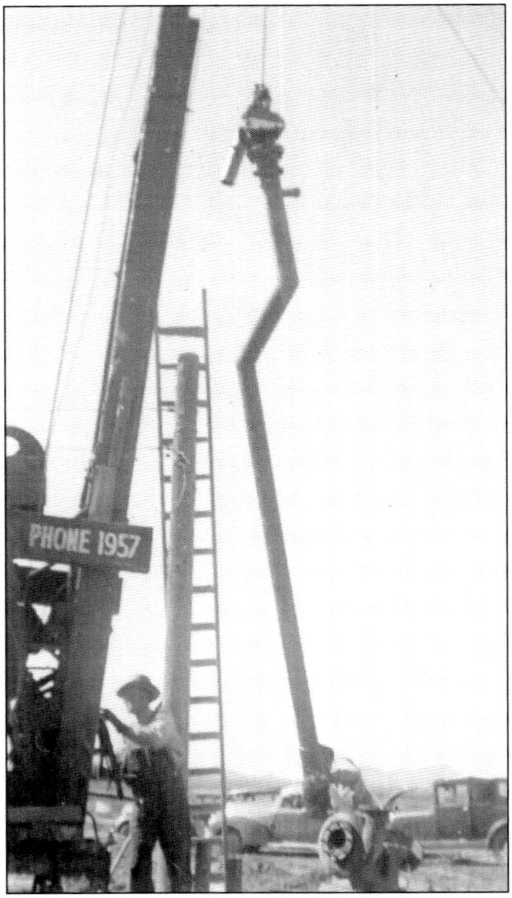

Around 1938, telephone lines and utilities were installed at Gilpin Airport. A 1930s Hudson is shown in the center of the photograph. Notice that the phone number on the pole is 1957. (Arizona Historical Society.)

Around 1938, workers had to climb the poles during construction for installation of electric and telephone services at Gilpin Airport. Reddington Pass can be seen in the background looking east. (Arizona Historical Society.)

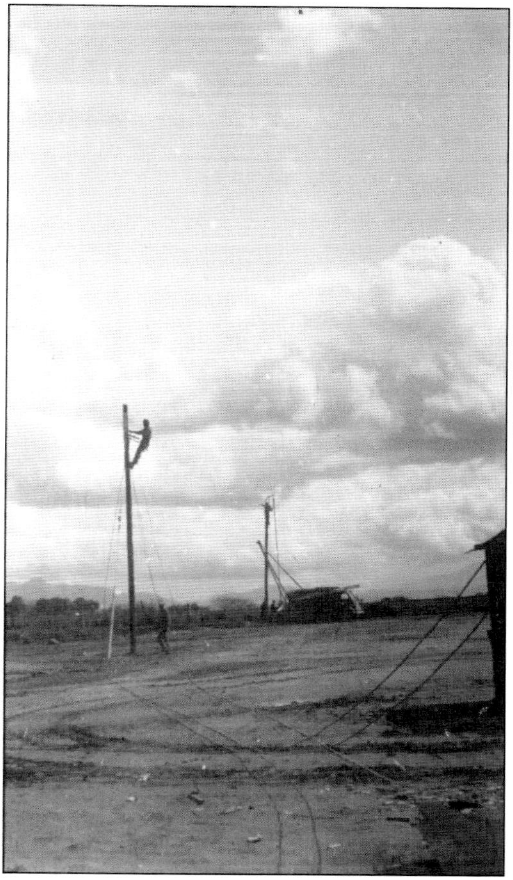

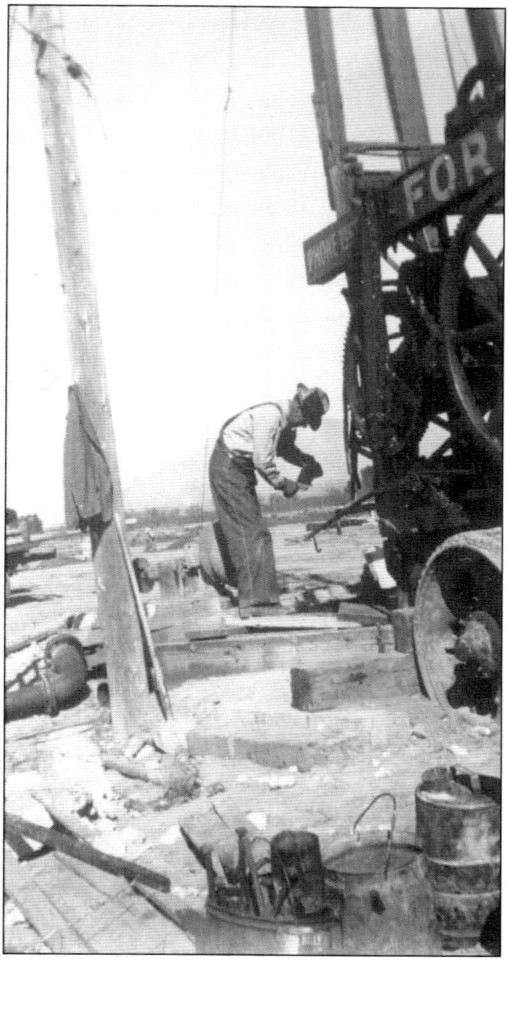

This worker is pictured in front of a rig believed to be for stringing and tensioning power lines in the early stages of the construction of Gilpin Airport. (Arizona Historical Society.)

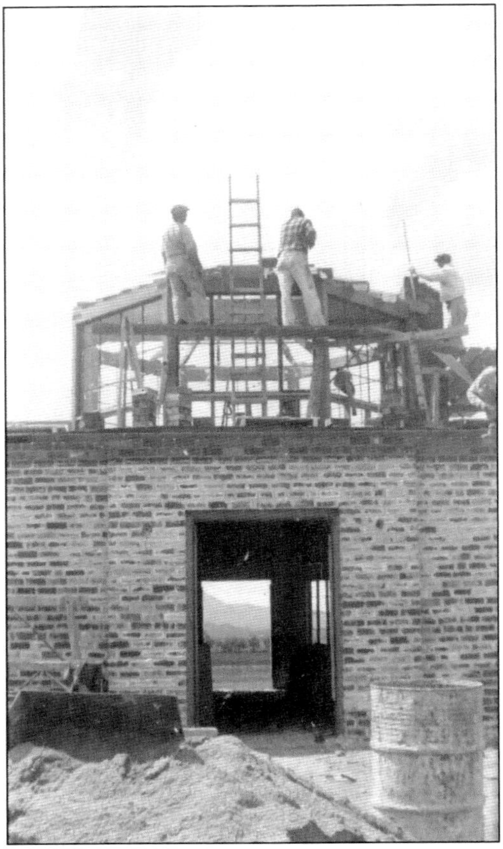

This is construction at Gilpin Airport around 1939. The administration building was between the two hangars on the eastern side of the field. This building included a lunch room, offices, a classroom for 20 students, the log room in which student flight hours were recorded, a pilots' lounge, a shower and bedrooms, and a central control tower where there was a weather bureau and lookout deck. (Arizona Historical Society.)

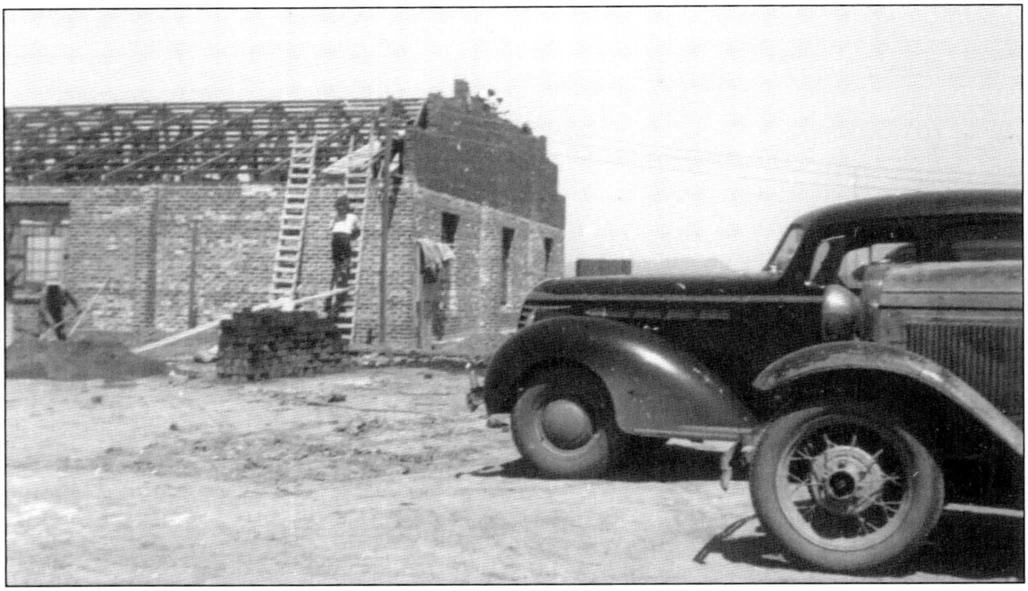

Construction started around 1938 on one of Gilpin Airport's new hangars. Built out of bricks, Gilpin opened in May 1941. One hangar was 70 by 100 feet, and the second was 90 by 100 feet and capable of storing 30 small planes. The cars at right are 1930s Hudson and Ford Model A sedans. (Arizona Historical Society.)

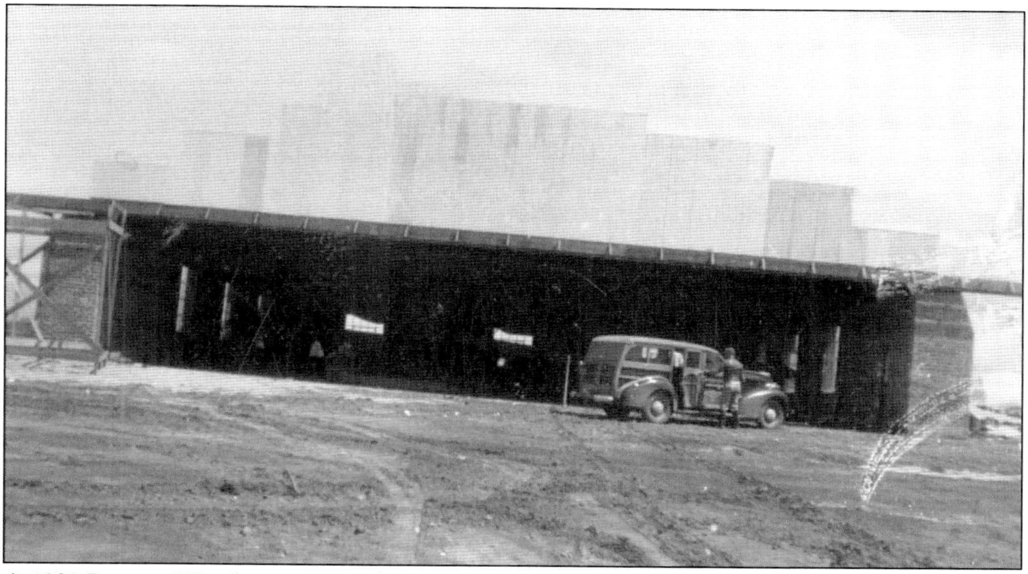

A 1939 Pontiac Woody transported passengers to and from Gilpin Airport; it is parked in front of a hangar still under construction. The tower building and two hangar buildings are still in existence today on Romero Road. The R.E. Darling Company operates at this location. (Arizona Historical Society.)

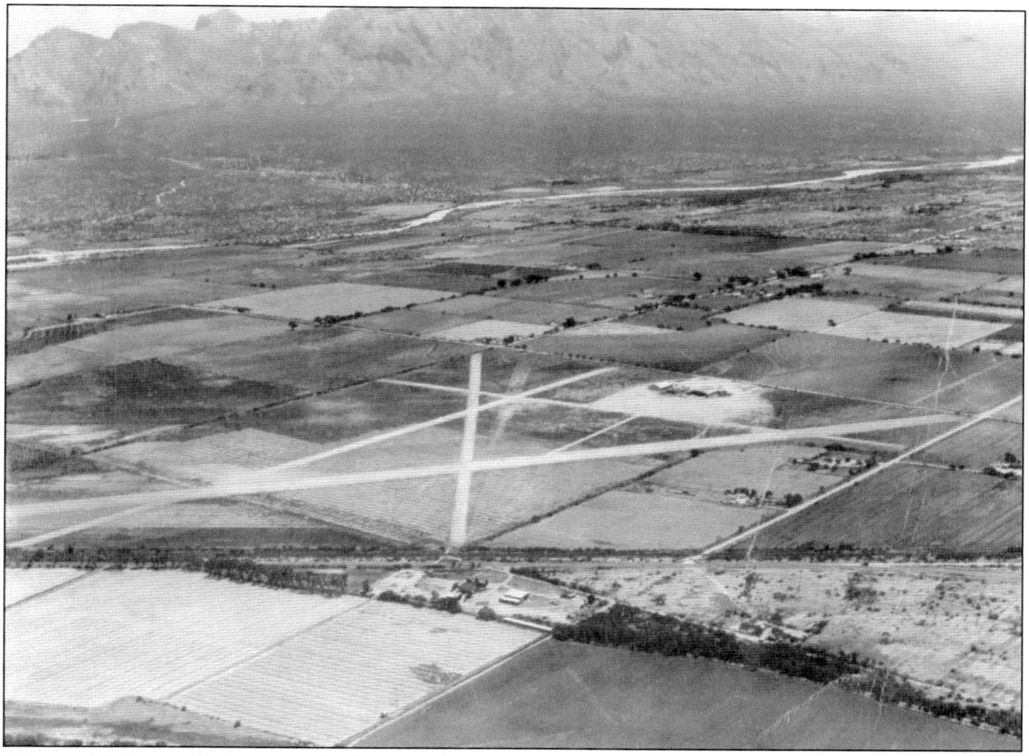

This c. 1941 aerial photograph shows Gilpin Airport between the Catalina Mountains in the background and cottonfields southwest of the railroad tracks. The runways (white X) are at center. Urban development changed the landscape after World War II. The intersection of Prince and Romero Roads is at the end of the large runway, on the far right. (Arizona Historical Society.)

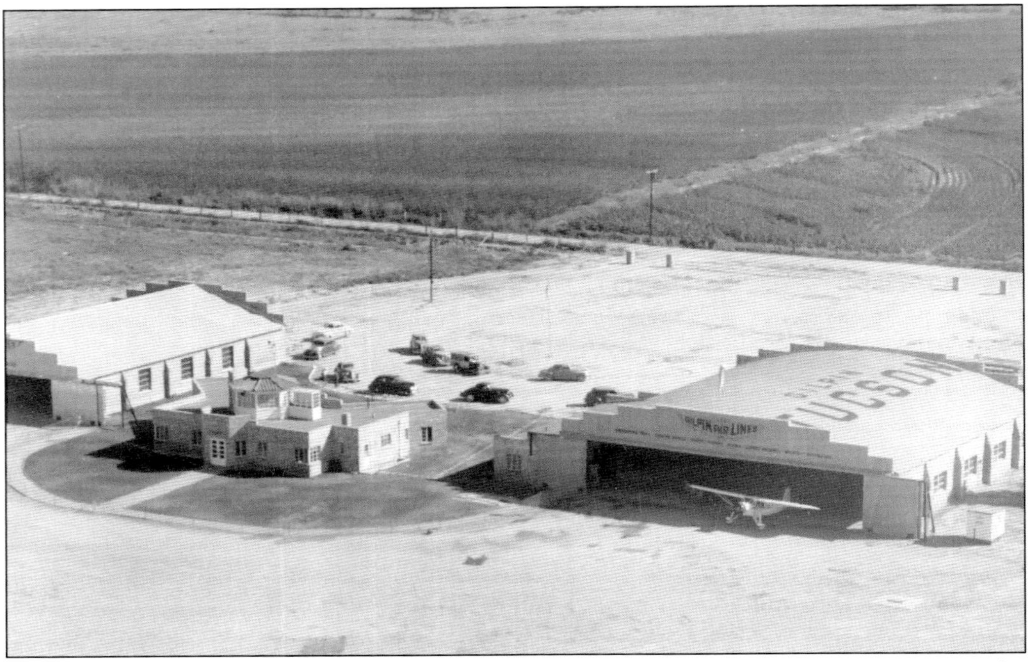

This is an aerial view of Gilpin Airport looking east, with Romero Road running across the top. Gilpin Airport covered 245 acres at the corner of Prince and Romero Roads. The property extended west to the Southern Pacific Railroad tracks and northwest, paralleling the railroad property for approximately one-half mile. (Arizona Historical Society.)

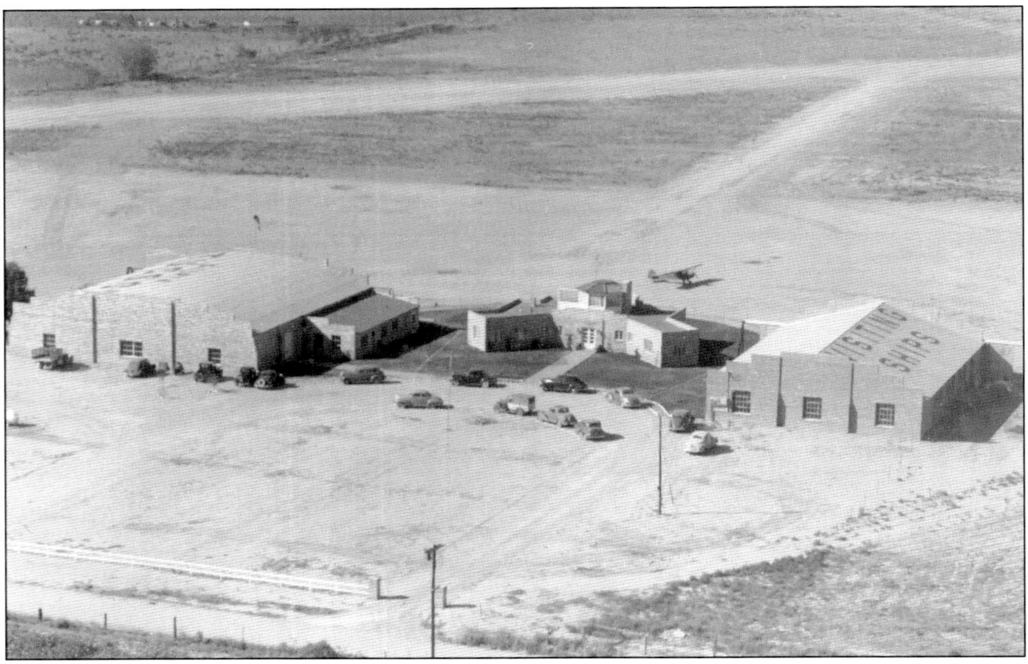

This is a c. 1941 aerial view of Gilpin Airport looking west, with Romero Road running along the bottom. Gilpin Airport consisted of two airplane hangars, shops, and an administration building. The buildings are still in existence today and are owned and occupied by the R.E. Darling Company. (Arizona Historical Society.)

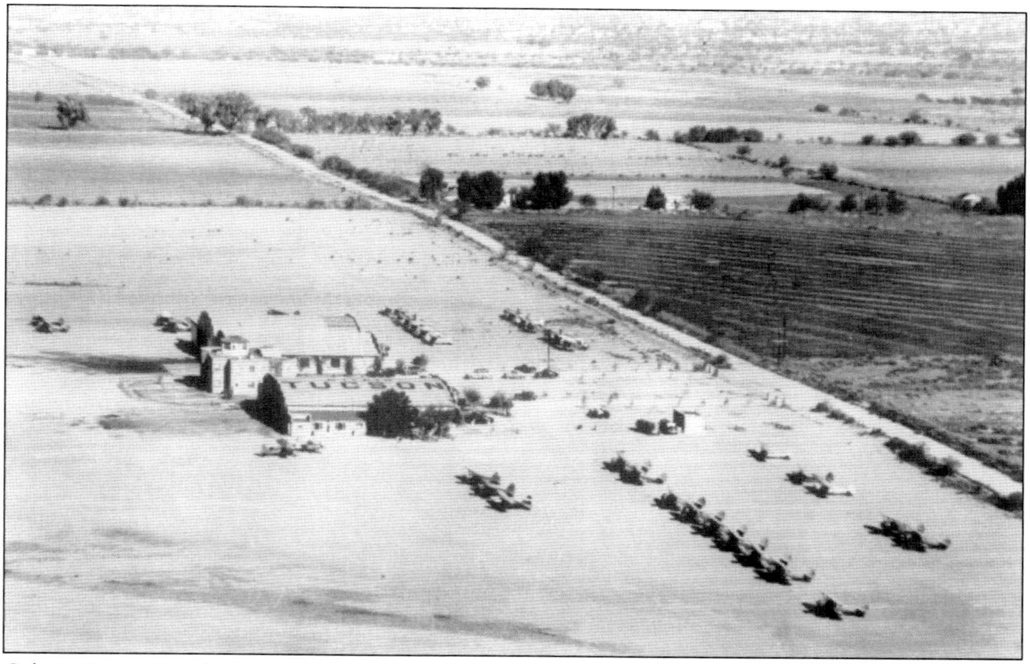

Gilpin Airport and the surrounding area are seen facing northeast, with Romero Road running diagonally at the top. Gardner Lane is at the northern boundary of the airport. G&G Airlines was founded by Charles W. Gilpin (killed in a plane crash near Mexico City in 1932) and Isabella Greenway (godmother to Walter Douglas Jr.). In 1929, Gilpin Airlines emerged in Glendale, California, with Greenway as a frequent passenger. The airline flew between Glendale and Ensenada, Mexico, but the lucrative tourist trade ended with Prohibition. Greenway moved the airport to Tucson in 1934 and asked her godson to move from San Francisco to run it. (Arizona Historical Society.)

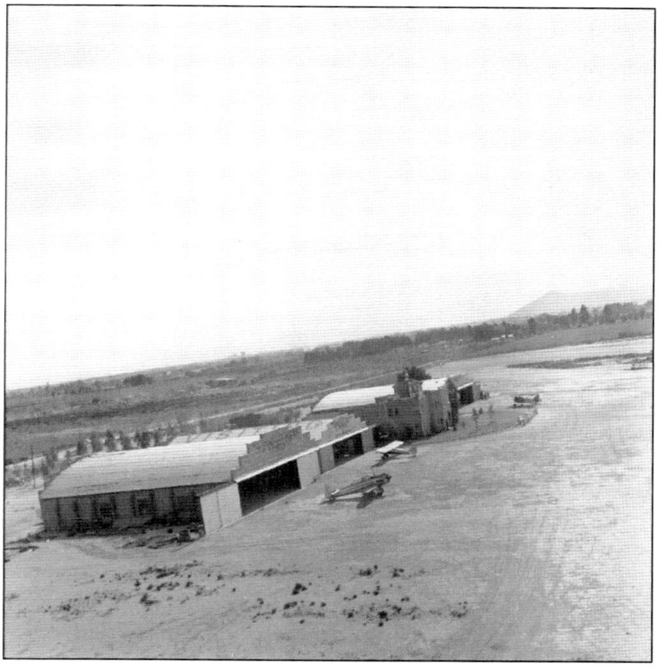

This c. 1941 view of Gilpin Airport faces south with Sentinel Peak in the background. Walter Douglas Jr. built a hangar and opened G&G Airlines at the Tucson Municipal Airport (now Davis-Monthan Air Force Base). He trained students and flew tours over the Grand Canyon through his subsidiary company, Grand Canyon Airlines. His equipment included an Aeronca C-3 "Flying Bathtub," used for student training, as well as open-cockpit Consolidated Fleet planes. When the Air Force base was created, Douglas opened Gilpin Field near Prince and Romero Roads. He sold the land in 1956. (Arizona Historical Society.)

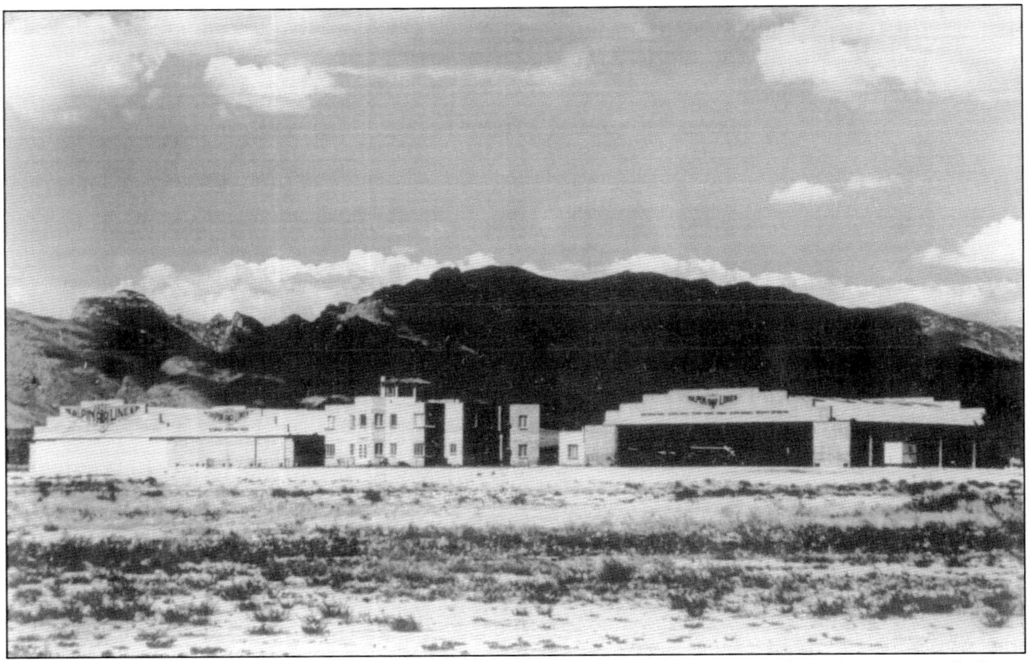

On a beautiful day in Tucson, this c. 1941 view of Gilpin Airport was taken facing north from the runway, with the Catalina Mountains in the background. Marcia Claridge remembers fondly the times when a blimp would land at the airport, which was quite a sight to see for the locals. (Jeretta Douglas.)

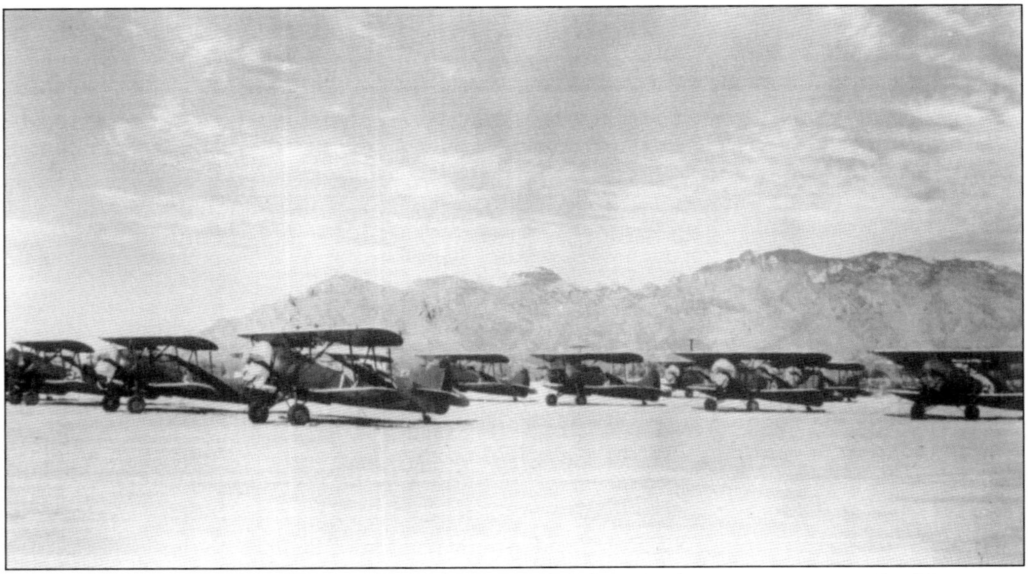

The Civil Aeronautics Authority (CAA) approved Gilpin Airlines as instructor for University of Arizona civilian fight training in 1939. Around 1942, Gilpin trained Army and Navy glider pilots. Pictured around 1941 at Gilpin Field, Taylorcraft, Waco UPF-7, and Fairchild PT-19 planes were used. The CAA closed its offices in Tucson in 1950, and programs were serviced out of Phoenix from that point. Only four CAA designees in Tucson were qualified to license engines and aircraft. They were Babe Clements, Gordon Hamilton, Al Hudgen, and Walter Douglas Jr. (Jeretta Douglas.)

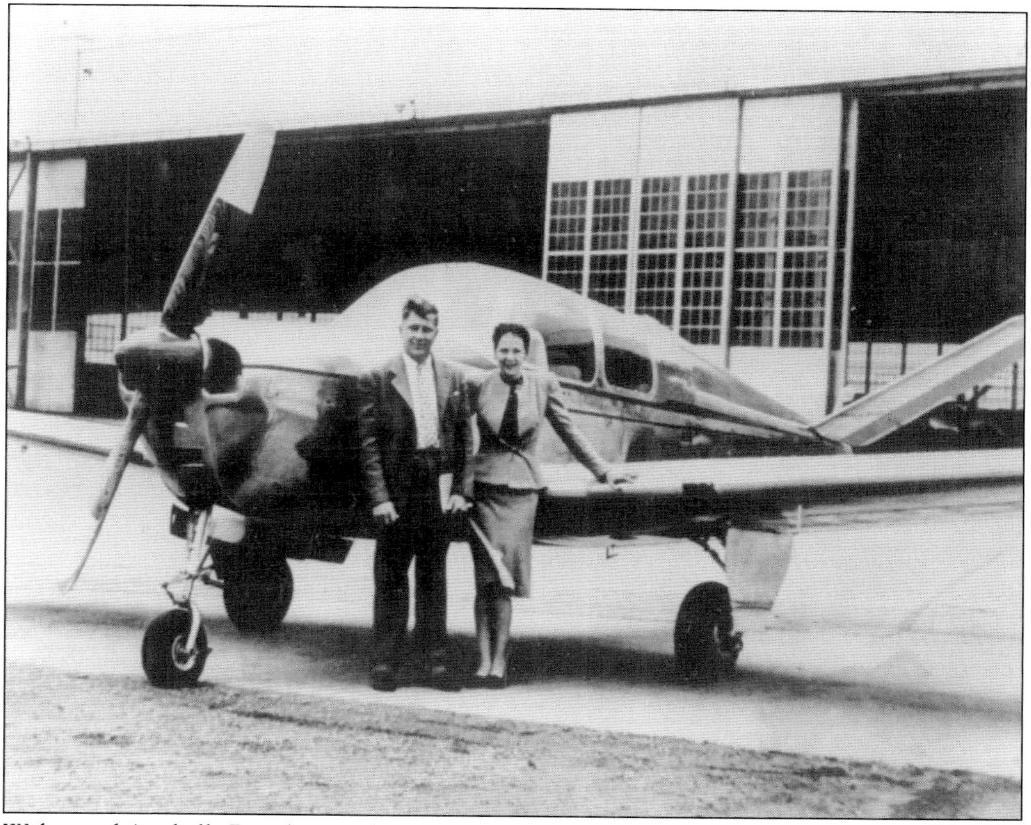

Walter and Anabelle Douglas stand in front of their 1948 Beach Bonanza at Gilpin Field. In 1938, Walter Douglas married Annabelle Spriggs-Wagner. Annabelle graduated from Tucson High and attended the University of Arizona. She was very active, helping at the airport, the racetrack, the Flowing Wells School District PTA, in the community, and at the University of Arizona. Their children attended Flowing Wells High School: Firelei (FWHS 1959), Walter III "Nibs" (FWHS 1960), Donald (FWHS 1962), and Elizabeth "Ellie" (FWHS 1966). (Jeretta Douglas.)

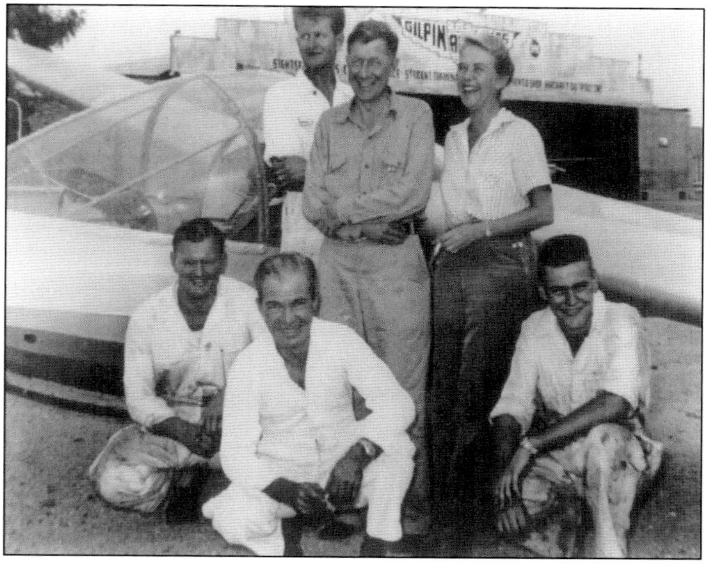

Walter Douglas, Anabelle Douglas (back row, right), and mechanics pose in front of a glider at Gilpin Field around 1942. (Jeretta Douglas.)

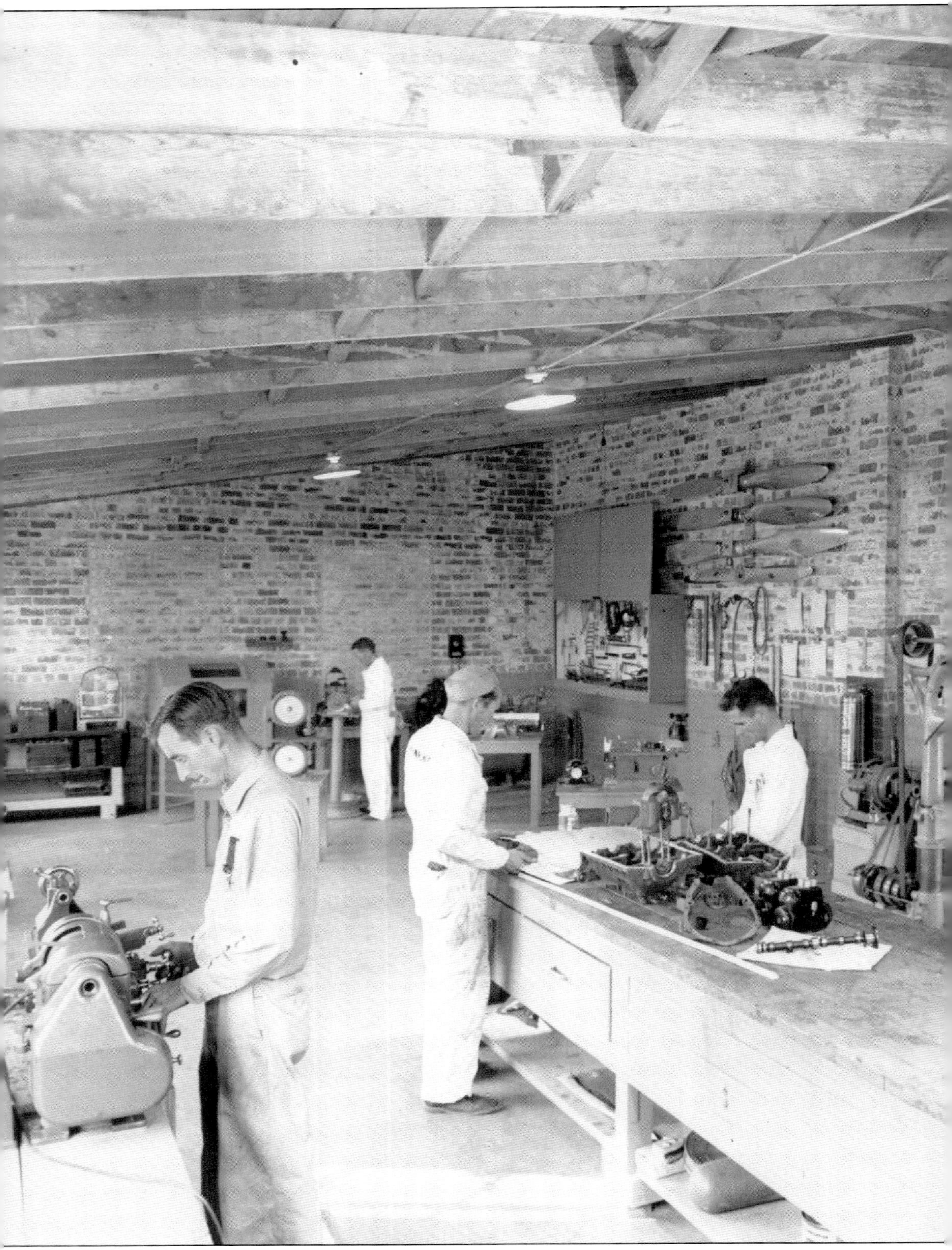

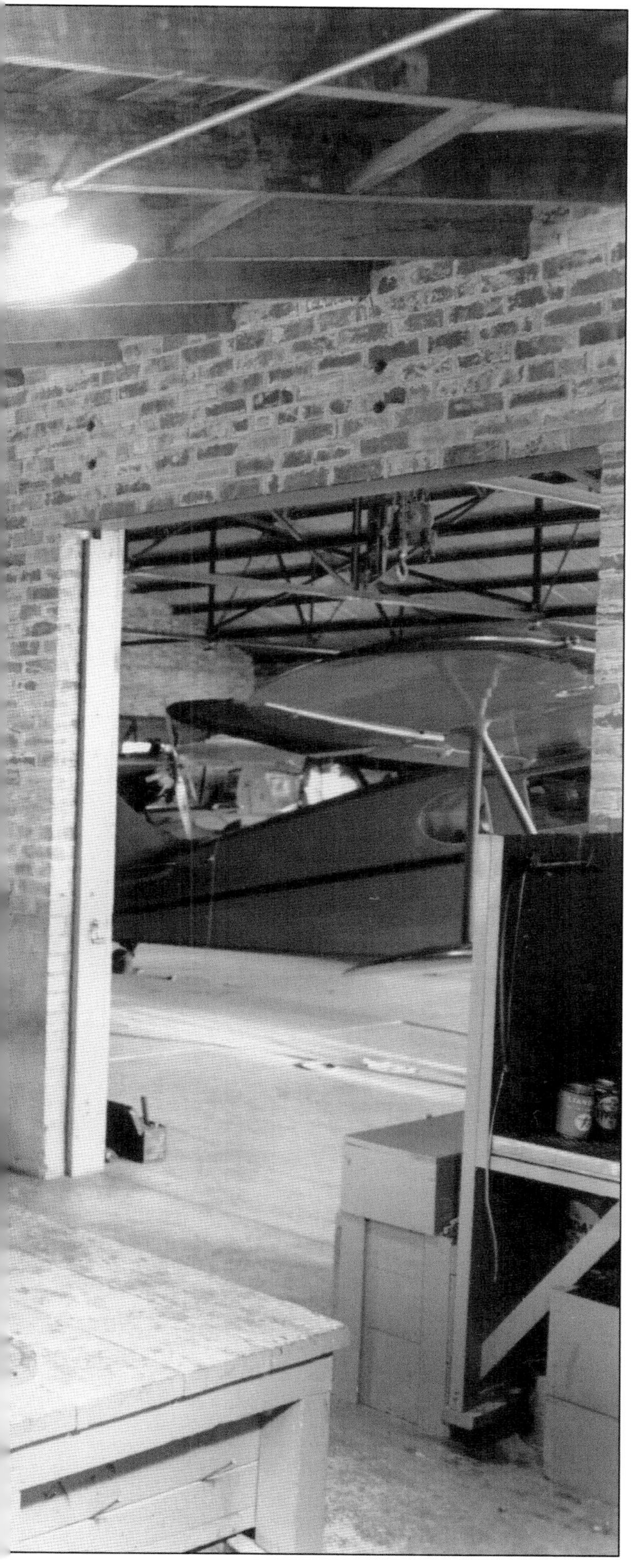

Workers at G&G Airlines tear down engines and rebuild them in the airport's machine shop around 1942. (Arizona Historical Society.)

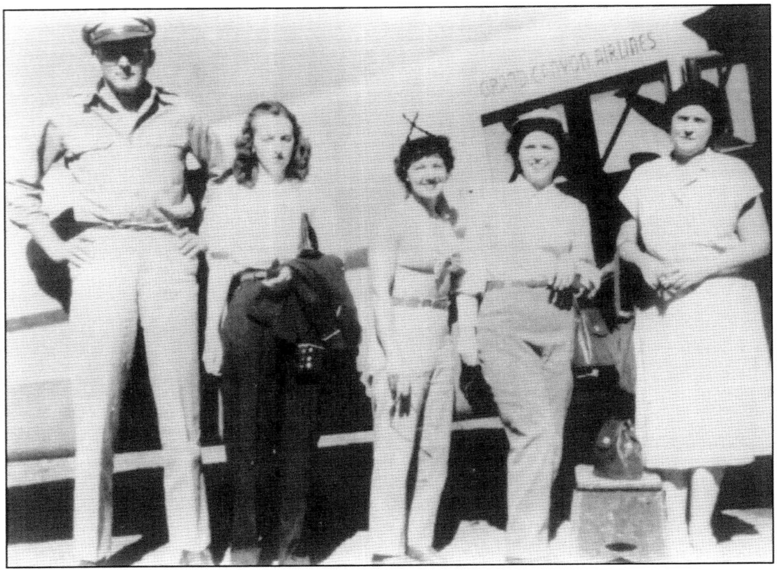

A pilot, crew, and passengers are pictured in front of Douglas's Grand Canyon Airlines airplane around 1940. (Jeretta Douglas.)

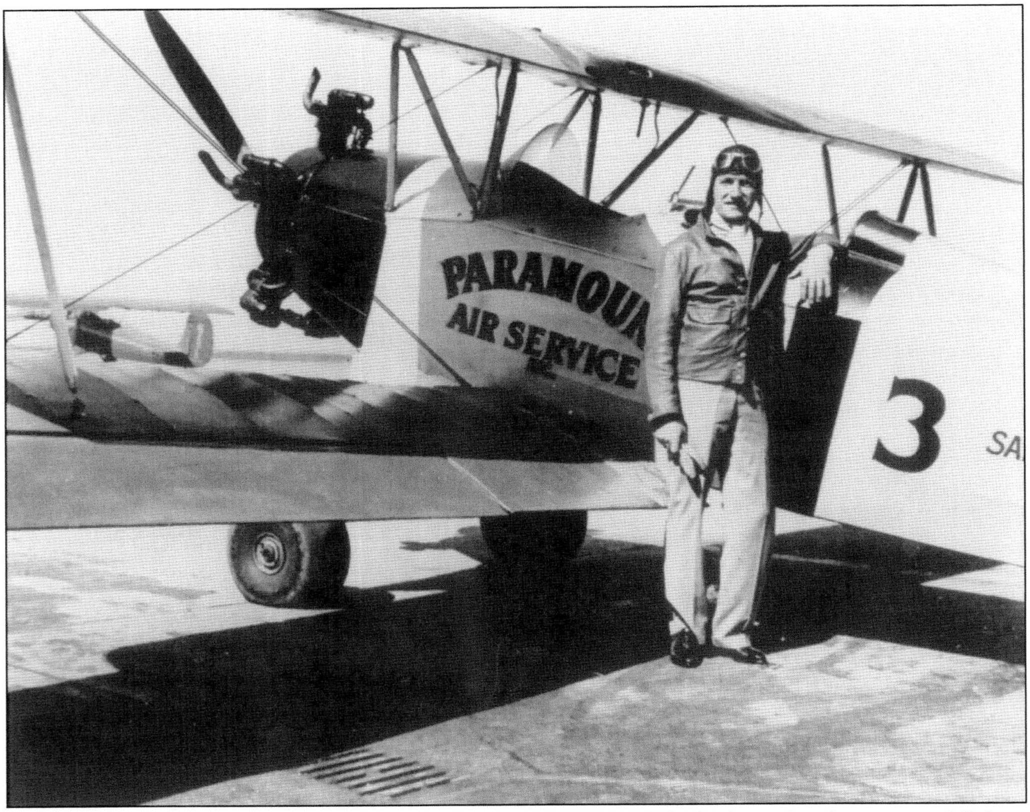

H.S. "Pete" Jacobs of Tucson poses in front of a 1929 Fleet Model 2 biplane. Jacobs managed Gilpin Airlines. Douglas learned how to fly in this plane in 1930. His instructor was Heinz Wegner. Douglas purchased Paramount Air Service and the Fleet. The airplane was brought to Tucson in 1934, and Heinz Wegner came too. Heinz later was part of the real estate firm Wegner and Associates, which ultimately purchased the Gilpin Airport land in 1956 and resold it to Abrams Airborne and other businesses. (Jeretta Douglas.)

Harold S. "Pete" Jacobs stands in front of a BT-13 aircraft around 1945. Pete and James W. Shields, an instructor for Gilpin, were killed instantly when their plane crashed into the side of a knoll about 10 miles northwest of Williams, Arizona, on February 28, 1948, in a heavy snowstorm. It took 15 days to locate the wreckage due to the storm. (Jeretta Douglas.)

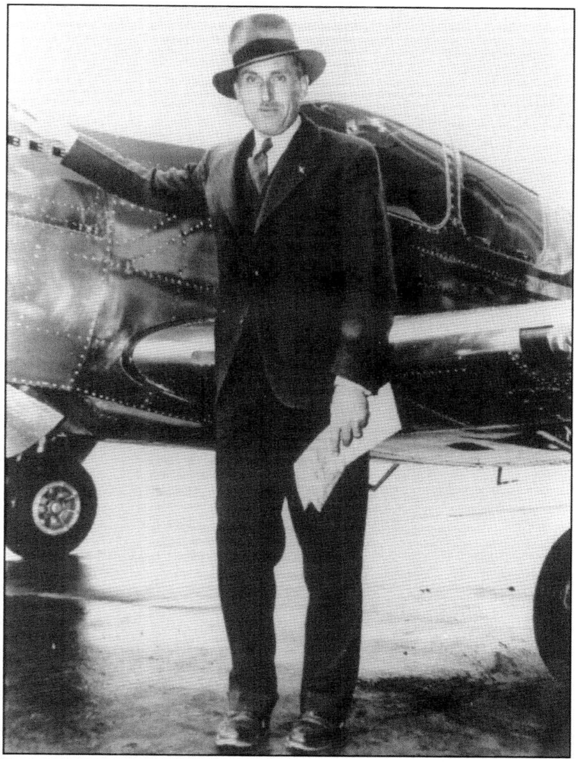

Members of a class of CAA pilot trainees are pictured at Gilpin Airport around 1945. The government contracted with Gilpin to provide instructional pilot training before, during, and after World War II. G&G Airlines trained over 1,000 pilots. (Jeretta Douglas.)

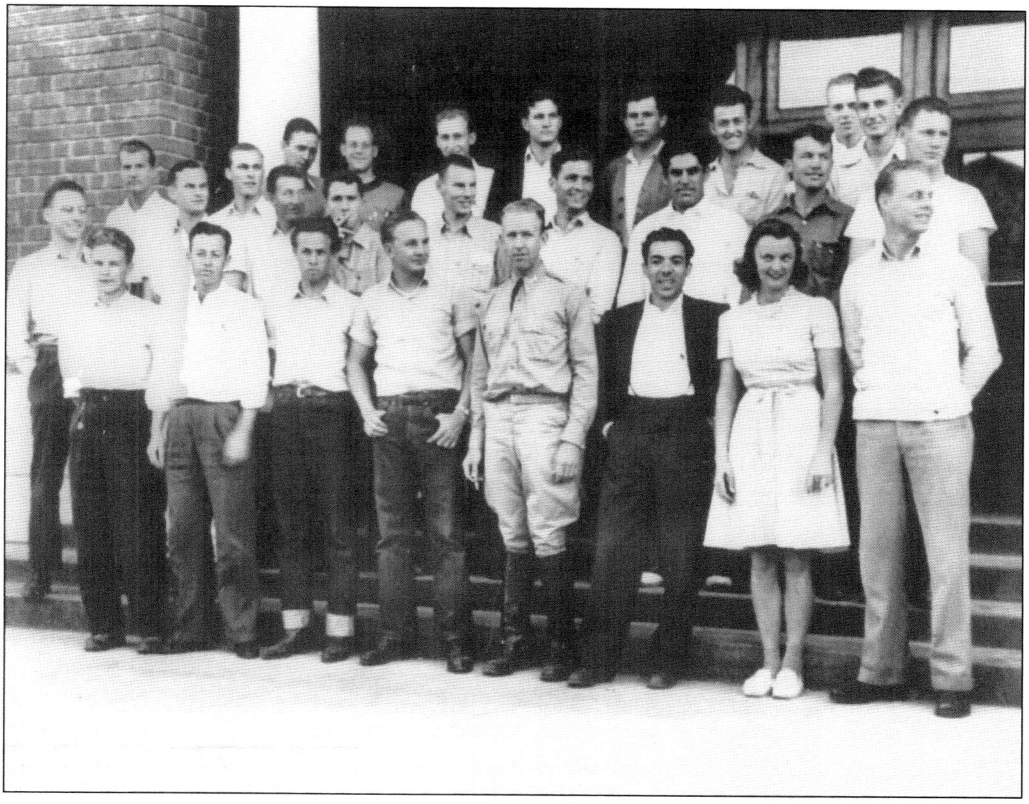

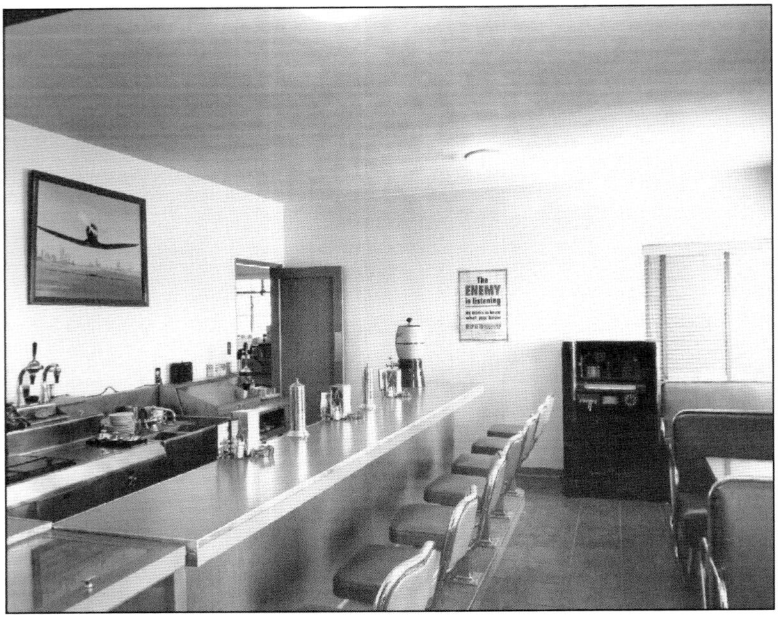

This was the café inside the Gilpin Airport around 1941. It served the typical café food of hamburgers and ice cream sundaes. The poster in the background warns that "The Enemy Is Listening." (Arizona Historical Society.)

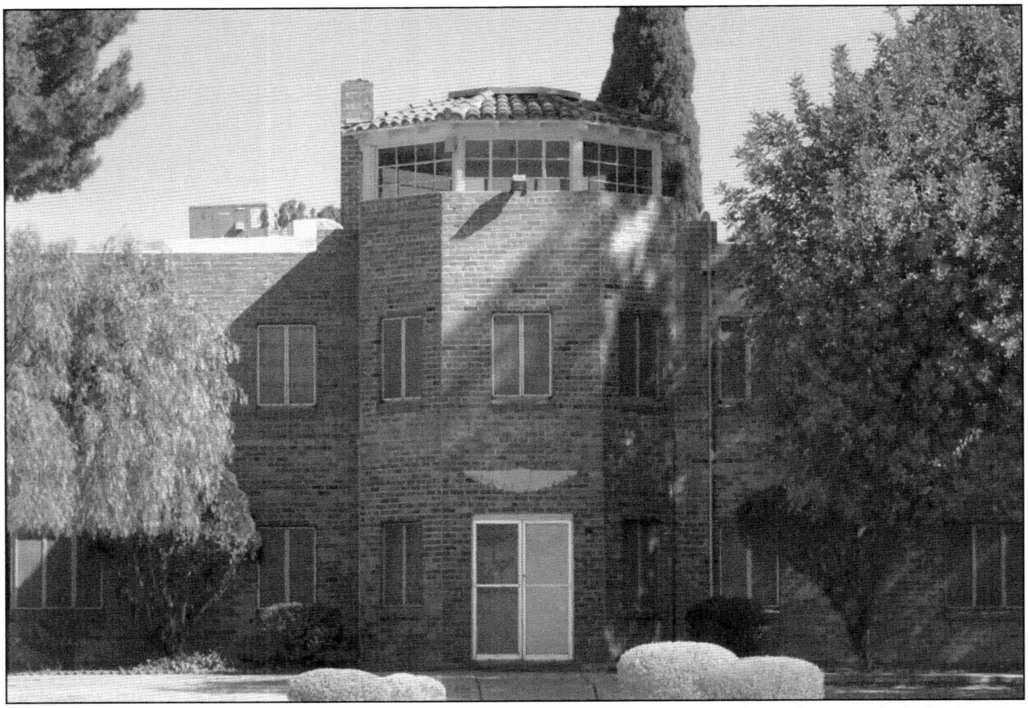

This 2018 photograph shows the Gilpin control tower that was purchased around 1957 for the home of the Tucson Aerospace Systems Division of R.E. Darling Co. (REDAR). REDAR began operations in 1948, fabricating high-altitude ventilation, communications, and wire hoses. Its life support hose and ducting were aboard the first Mercury flight into space. REDAR made much of the life support equipment for the Mercury, Gemini, and Apollo programs and for use in the lunar excursion module and lunar backpacks. R.E. Darling Co. also built the breathing apparatus for the Skylab ergometer, which was used for collecting breath samples in space, along with providing breathing hoses used in all of the space shuttle flights. (R.E. Darling Co. and authors.)

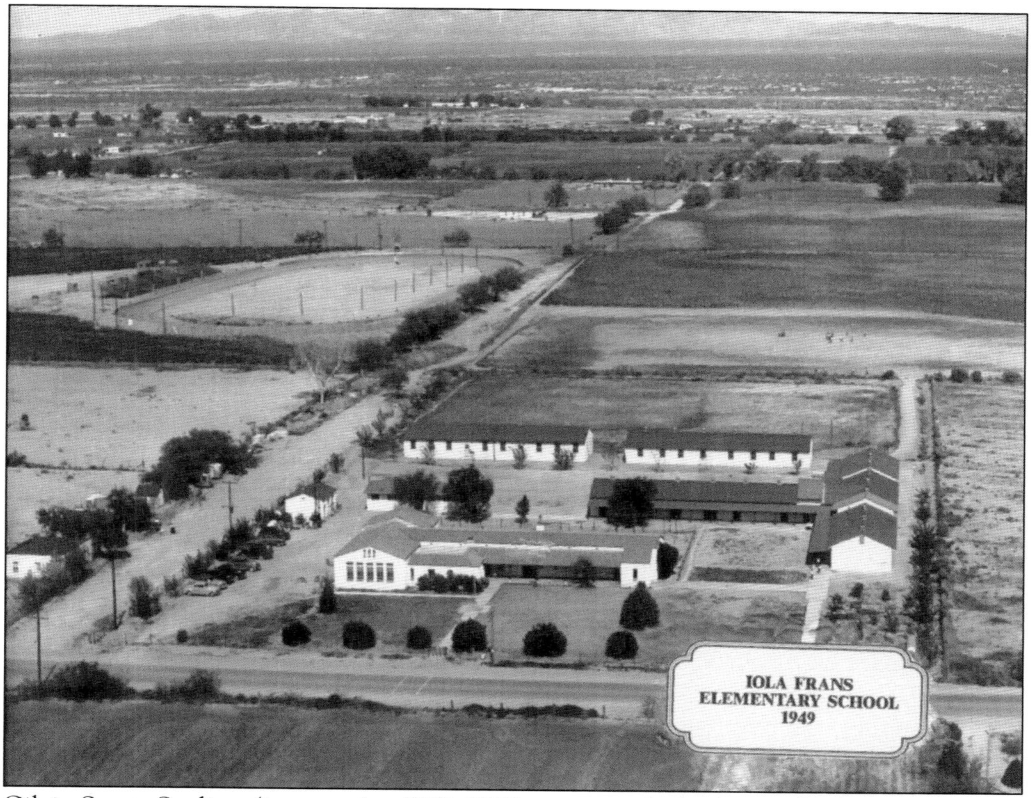

Gilpin Sports Stadium (in operation 1949–1952) is seen in the background behind the Flowing Wells School in 1949 (renamed Iola Frans Elementary School in 1956). At a cost of $62,000, the stadium hosted stock car, midget car, motorcycle, and jalopy races. The seating capacity was 2,500. Water Douglas was the engineer of the stadium's track. It included a snack bar, modern bathrooms, and lighting for racing at night. The Gilpin Stadium was directly west of where the current FWHS track is located. (FWUSD.)

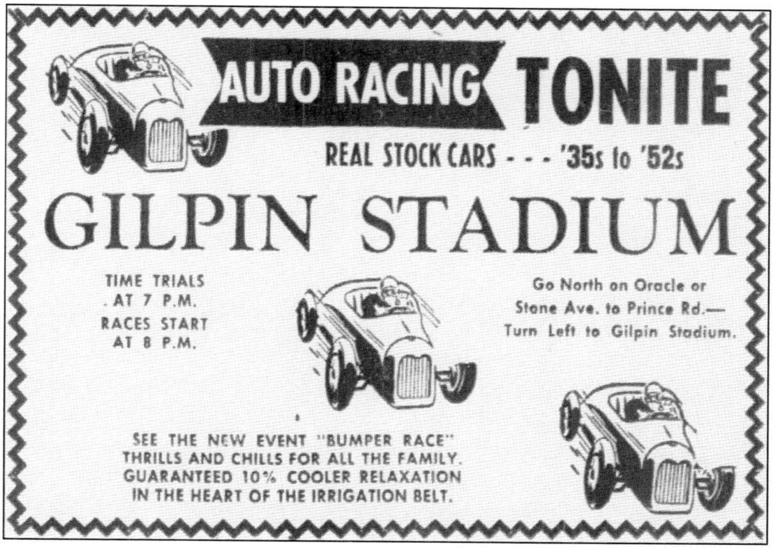

This is an advertisement for events at Gilpin Stadium in 1952. Gilpin Stadium hosted local favorites Bill Cheesbourg and Roger McCluskey. McCluskey, a Tucson High graduate, was an 18-time veteran of the Indy 500 and had four top-10 finishes. He finished third in the Indy 500 in 1962 and again in 1973. (*Arizona Daily Star.*)

Record Is Broken At Gilpin Stadium

Three drivers fractured the same track record at Gilpin stadium yesterday during motorcycle and jalopy racing events. Wild Bill Cheesbourgh's 22.75 mark registered last Sunday was bettered by Frank Heaton's mark of 22.61, followed by Bob Stewart's 22.60.. and finally, Kurt Steinbach's 21.70, which is now the record.

The jalopy driven by Kessler Mitchell crashed through the safety fence in heat No. 1. He was not injured.

The 30-lap main event provided the best racing seen thus far this season while drivers narrowly missed crashing into the retaining wall or into spinning jalopies.

Complete results of yesterday's races follow:

Motorcycle trophy dash—3 laps. 1—Musselman 57.81, 2—Bob Stewart.

Jalopy trophy dash—3 laps. 1—Steinbach 1:12.04, 2—Bob Stewart.

Jalopy heat No. 1—8 laps. 1—Rhinehart 3:16.80, 2—Red Eiland, 3—Bob Ison.

Jalopy heat No. 2—8 laps. 1—Martin 3:03.98, 2—Cheesbourg, 3—Steinbach.

Motorcycle consolation—6 laps. 1—Hickman 2:01.90, 2—Sisler, 3—Gibson.

Motorcycle main — 15 laps. 1 — Brenton 4:58.50, 2—Musselman, 3—Jean.

Jalopy main — 30 laps. 1 — Steinbach 11:30.40, 2—Cheesbourg, 3—Thrope.

Motorcycle two-up race—6 laps. 1—Musselman and Froy 1:58.60, 2—Stewart and Johnson.

Jalopy match race—5 laps. 1—Cheesbourg 1:57.00, 2—Johnny Dye.

"Wild" Bill Cheesbourg was born in 1927. This Tucson High graduate won his first soapbox derby on A Mountain at the age of 11. He built and raced jalopies during the 1950s at the Gilpin Stadium and other dirt tracks in the area. He raced in seven Indy 500s and was inducted into the Tucson Raceway Hall of Fame's first class in 1992. (*Arizona Daily Star*.)

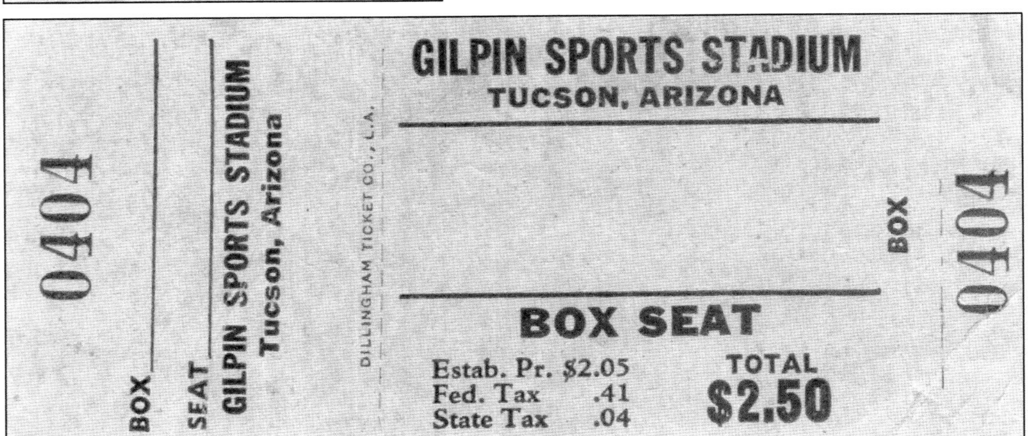

This is a ticket from Gilpin Stadium. Local residents definitely noticed the noise and dust from the racetrack. A box seat ticket to watch the races on the quarter-mile dirt track cost $2.50. Bus service to the track was provided to and from Congress Street and Stone Avenue in 1949. (Jeretta Douglas.)

Five

The Dairies and Businesses

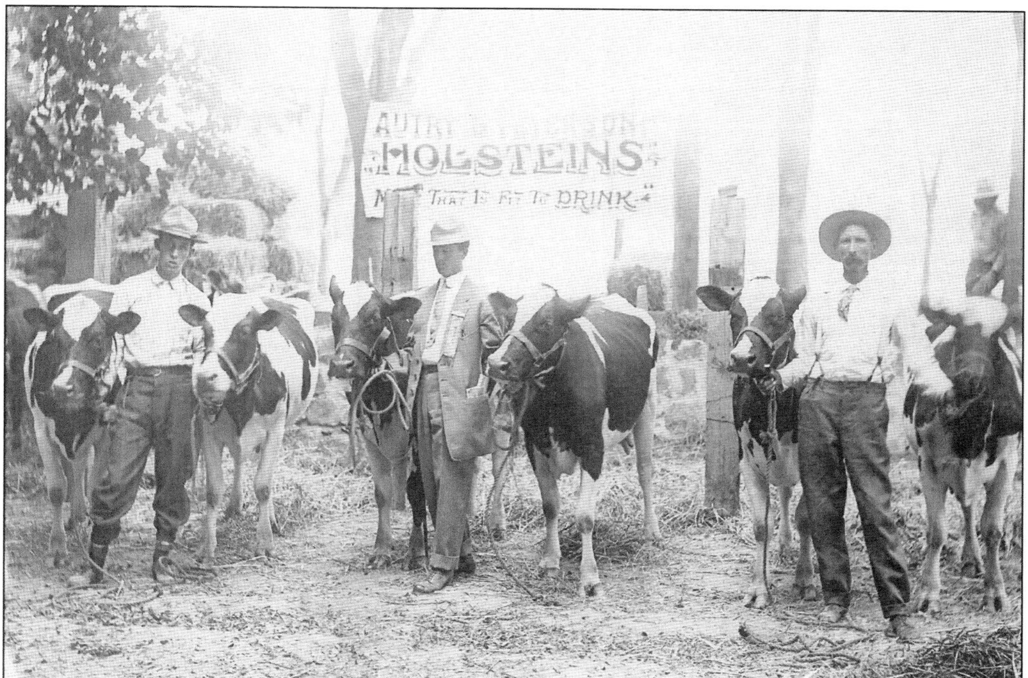

This picture was taken on October 26, 1911, a rainy day at the first Pima County Fair (October 25–27). There were 779 entries, not counting educational exhibits. At the cattle judging, Marion Autry and John Peterson won first and second place for "nice grade Holstein Cows," according to the *Arizona Daily Star.* Flowing Wells Dairy won first "on a good registered Jersey bull, and on grade Guernsey and Jersey cows." Pictured from left to right are John Peterson; Prof. W.L. Fowler, University of Arizona College of Agriculture; and Marion S. Autry. In the early 1900s, Autry partnered with Peterson Dairy. Around 1923, Autry opened his own dairy with 20 Holstein, seven Jersey, one Ayrshire, and two Guernsey cows, naming it the Modern Dairy. (Diane Peterson Graham and Maynard "Pete" H. Peterson.)

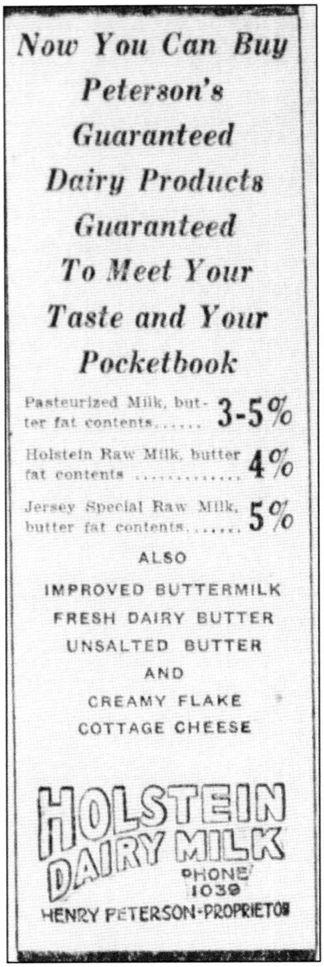

This advertisement for the Holstein Dairy appeared in the *Arizona Daily Star* on January 27, 1932. Milk was sold by the amount of butter fat it contained. (*Arizona Daily Star*.)

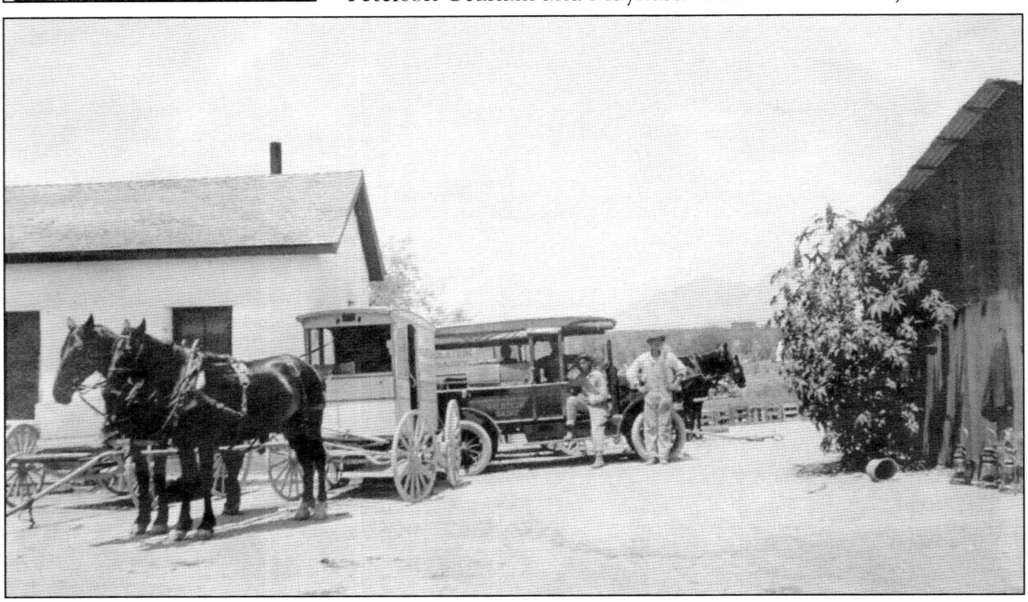

Peterson Dairy milk delivery wagons were both horse-drawn and motorized in the 1920s. Here is a 1920s International delivery wagon. The Drachman Street location was believed to be their offices and distribution center. The western edge of the Catalina Mountains can be seen in the background to the north. (Diane Peterson Graham and Maynard "Pete" H. Peterson.)

This is believed to be the Peterson brothers' Holstein Dairy building on Drachman Street. The 1920s Dodge truck was used for milk deliveries. (Diane Peterson Graham and Maynard "Pete" H. Peterson.)

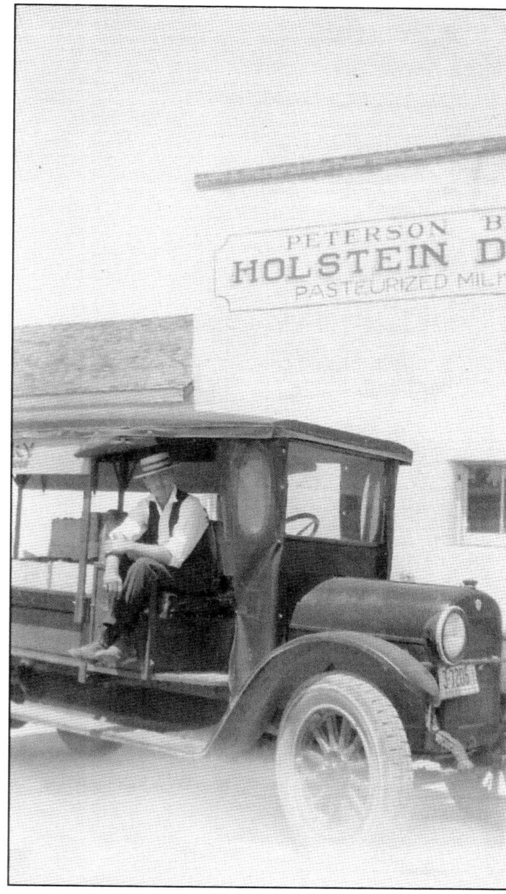

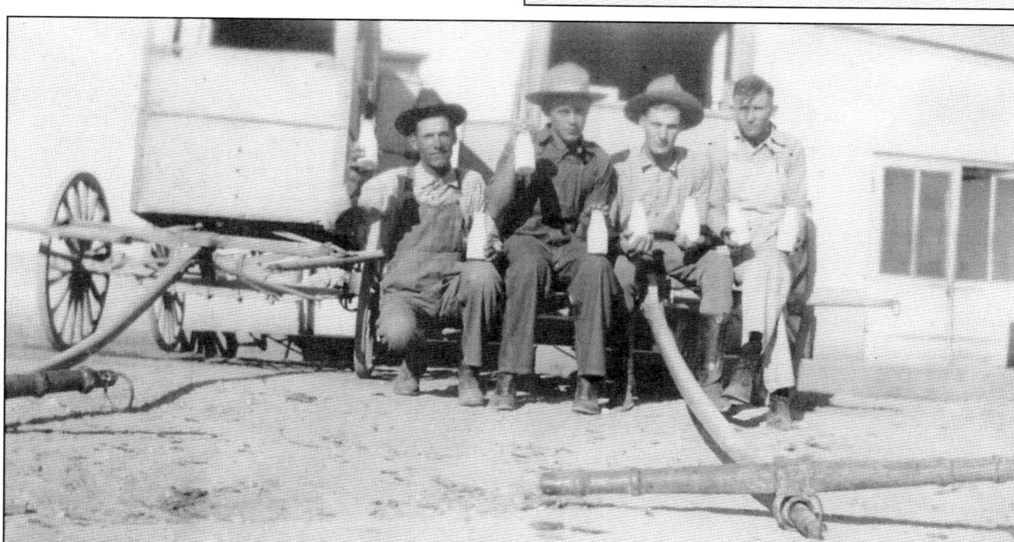

Workers at the Peterson brothers' Holstein Dairy hold bottles of milk next to a 1920s milk delivery wagon. The Peterson farm was on Prince Road near the old Casa Grande Highway, State Route 84 (now Interstate 10). (Diane Peterson Graham and Maynard "Pete" H. Peterson.)

Saturday July 1st

FOX

10 a. m. Sharp

Look Kiddies Look!!

Big Musical Vodvil Act

"The Alabam Boys"

Election of Club Officers

25 Free Passes to Mission Swimming Pool

10 Stamp Collection Cards, Dooley Birthday Gift Cake—Puhl's Bakery

5th Chapter Serial, "Hurricane Express"

Feature, "Little Orphan Annie"

Chocolate Milk for Every Child By Peterson's Dairy

This advertisement appeared in the *Arizona Daily Star* on Saturday, July 1, 1933. The dairy advertised chocolate milk for every child at popular movies and performances at the Fox Theater downtown. (*Arizona Daily Star.*)

FAIRVIEW DAIRY LEADS LIST IN MILK SCORES FOR JANUARY

Holstein Dairy Products Lead in Score for Cream for Same Period.

Tests of the milk disposed of by every dairy in this locality and firms handling milk products within the city during the month of January disclosed that in only one instance was the bacteria content higher than that permissible by law. The full report of milk tests for the month of January was disclosed yesterday by City Health Officer Dr. A. G. \Schnabel.

Samples of milk from each dealer were taken near the end of January by Deputy Health Officer J. E. Dufton, tests being made by the laboratories at the . University of Arizona.

Requirements of the municipal milk ordinance are that milk have a maximum of 150,000 bacteria per cubic centimeter asu not less than 3.25 per cent of butter fat content. Scores apportioned by the city health officer are based upon butter fat and bacteria contents, 100 representin ga perfect score, according to the officer.

The Fairview and Jaynes dairies lead all others during the month in purity, both of these being found to have sold milk with a bacteria content of 3,000 per cubic centimeter. Gerhart's dairy, with a bacteria content of only 4,000, leads in butter fat contents with 4 per cent.

In cream products, the Holstein dairy leads in purity with 80,000 bacteria, while Marigold products lead in richness with 55 per cent of butter fat. Requirements of both bacteria and butter fat contents for cream are scaled considerably higher than for milk.,

The Fairview dairy leads in score for milk while the Holstein dairy leads in the score for cream. The January report follows:

Dairy—	Bacteria per c. c.	Milk Fat	Score
Fairview	3,000	3.4	94.9
Gerhart's	4,000	4.0	93.6
Jaynes	3,000	3.6	90.6
Riverside	15,000	3.8	90.0
Paradise	17,000	3.4	88.9
Hamrock	20,000	4.1	88.9
West End	11,000	3.8	87.7
Marigold	27,000	3.8	87.5
Clover Bar	16,000	3.4	86.0
Royal	55,000	3.6	83.1
Sunset	7,000	3.0	81.2
Modern	60,000	3.4	78.9
Yale Bulgarian ..	33,000	3.0	69.9
Holstein	57,000	3.0	66.4
Price Rillito	90,000	3.2	65.9
Purity	220,000	0.0	52.9
Cream Scores for January			
Holstein	80,000	18.0	76.7
Marigold	550,000	55.0	59.9
Royal	3,500,000	32.0	59.4
Sunset	3,100,000	35.0	57.4

This article appeared in the *Arizona Daily Star* on February 11, 1923, detailing milk scores of the time. Milk scoring and reporting of cream content, butterfat, and bacteria were regularly and prominently reported in the newspaper. (*Arizona Daily Star.*)

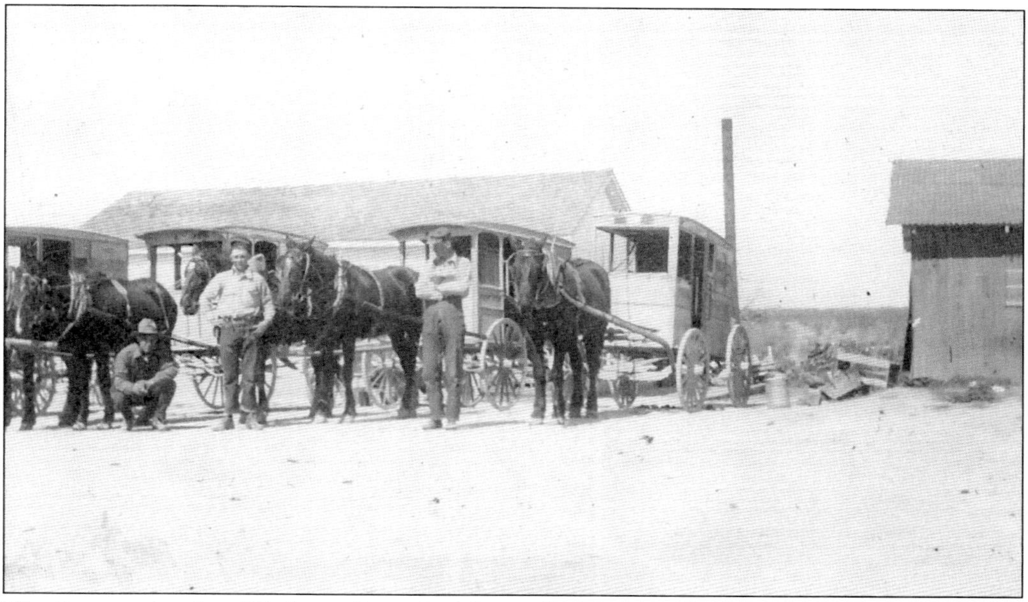

The Peterson brothers' Holstein Dairy had five milk delivery wagons to deliver 3,000 bottles of milk daily all over Tucson. The Peterson brothers received milk from six ranches with 250 cows that were milked twice daily. One Peterson ranch is believed to be at Sentinel Peak near what is now Mission and Silverlake Roads, and another was on Prince Road near what is now Interstate 10, with the main processing hub on Drachman Street near Stone Avenue. (Diane Peterson Graham and Maynard "Pete" H. Peterson.)

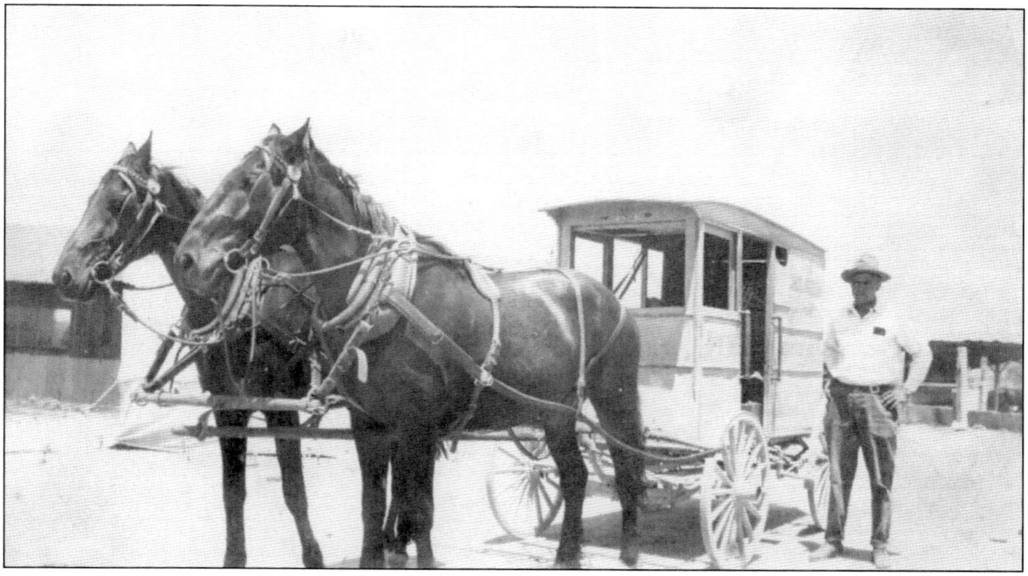

Henry Peterson of the Holstein Dairy is pictured with a milk delivery wagon in the 1920s. Peterson started the dairy in 1904. He married Mamie Bouschet, who was raised in the Rillito School District, in 1921 and raised six children. Bouschet and Harris family members were the early school board for the Rillito School District in the 1890s. Mamie was one of eight children in the class around 1896. Henry passed away on December 8, 1935. To honor him, Flowing Wells School was closed at noon on the day of his funeral in 1935. (Diane Peterson Graham and Maynard "Pete" H. Peterson.)

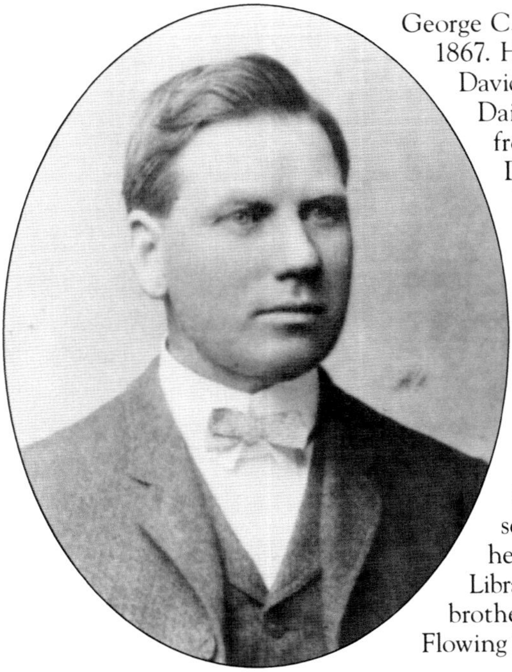

George C. Woods was born in Missouri on October 17, 1867. He came to Tucson in 1916 via Texas and St. David–Benson, Arizona, and took over the Modern Dairy. In April 1919, Woods purchased 144 acres from Tucson Warehouse and Transfer north of DeMoss-Petrie Road (Grant Road) and what is now Fairview Road. There he established the Fairview Dairy. His home was built by contractor Wylie E. Rudasil (for whom Rudasil Road is named). Fairview Road is named after this dairy. Upon Woods's death in 1932, his wife, Bessie, and sons George F. and William L. continued to run the dairy. They sold a portion of the land in 1951 to Tucson Light and Power Co. After the dairy closed in 1956, a portion of the land was sold to Pima County for the Interstate 10 right-of-way. Bessie passed away in 1967. George F. Woods served on the city council from 1961 to 1969, when he died suddenly. The Woods Memorial Branch Library on First Avenue is named in his honor. Younger brother William Leonard Woods later served on the Flowing Wells School Board. (Jean Woods O'Hanlon.)

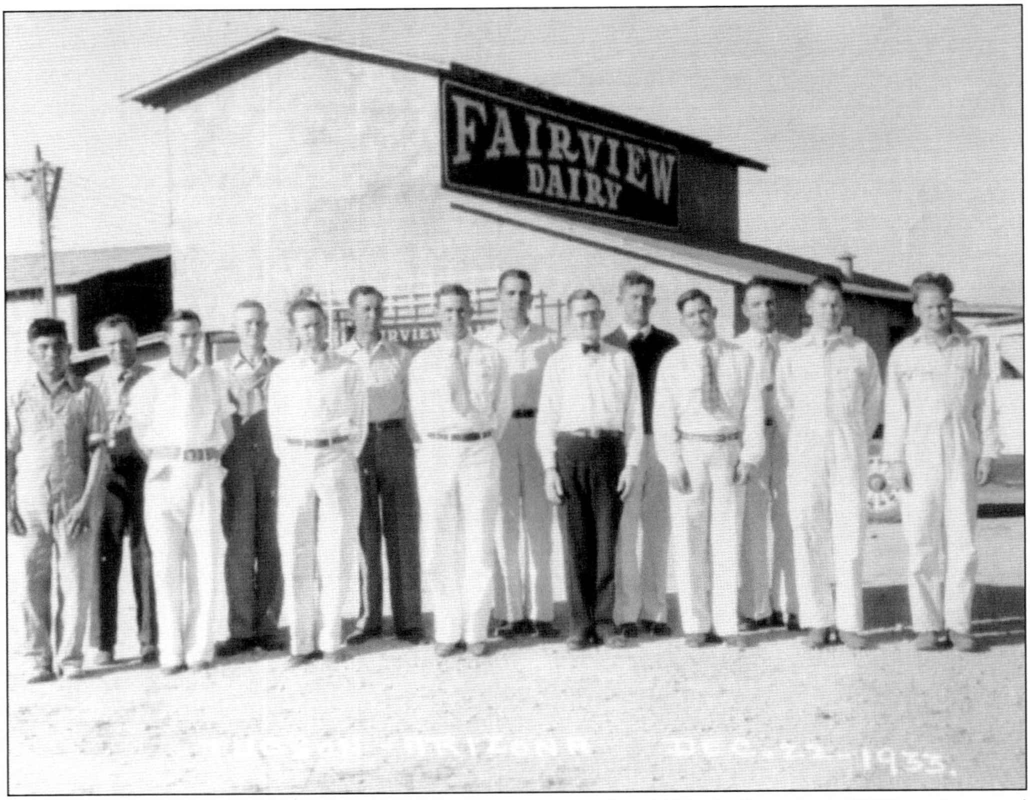

Fairview Dairy employees and the owner's sons are pictured in 1933. At far right is William Leonard Woods, and at far left is George Freeman Woods. After the dairy closed, the Farmer John's meat packing plant took over that location and remained there until 2001. (Jean Woods O'Hanlon.)

This advertisement for the Fairview Dairy was run in the *Arizona Daily Star* on February 27, 1938, featuring Ruth White, a nationally known home economist for the National Livestock and Meat Board. She was also the director of the Mansfield News-Journal Cooking School. She prepared food in lectures at the Ohio Theater in Mansfield, Ohio. Fairview Dairy was a featured product. (*Arizona Daily Star.*)

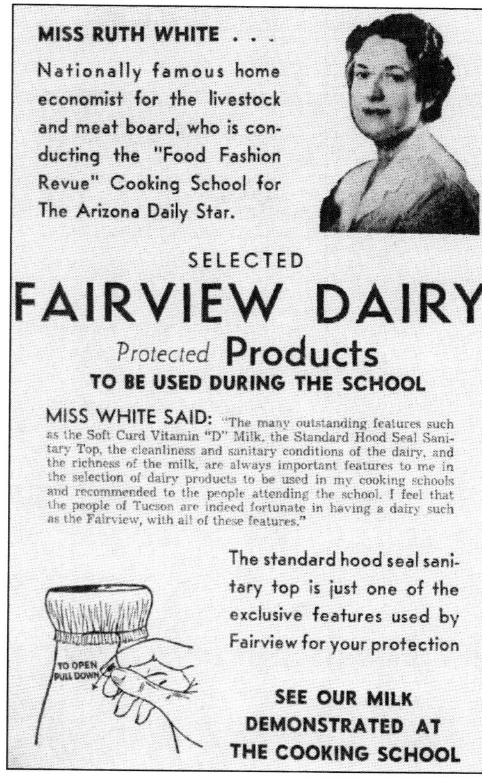

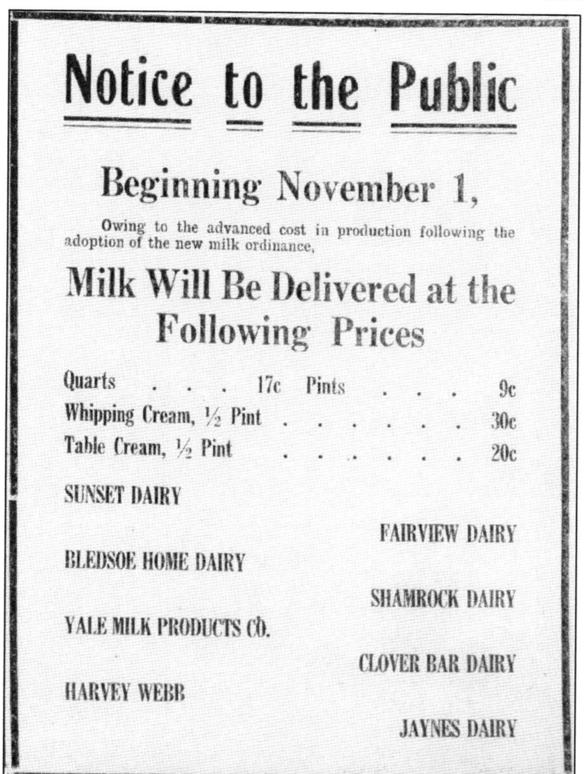

Notice to the Public

Beginning November 1,

Owing to the advanced cost in production following the adoption of the new milk ordinance,

Milk Will Be Delivered at the Following Prices

Quarts . . . 17c Pints . . . 9c
Whipping Cream, ½ Pint 30c
Table Cream, ½ Pint 20c

SUNSET DAIRY
FAIRVIEW DAIRY
BLEDSOE HOME DAIRY
SHAMROCK DAIRY
YALE MILK PRODUCTS CO.
CLOVER BAR DAIRY
HARVEY WEBB
JAYNES DAIRY

This notice in the October 30, 1925, *Arizona Daily Star* notified customers of an increase in milk prices due to the Tucson City Council requiring pasteurization of all milk. In September 1925, the city board of health warned housewives to boil their milk and water. (*Arizona Daily Star.*)

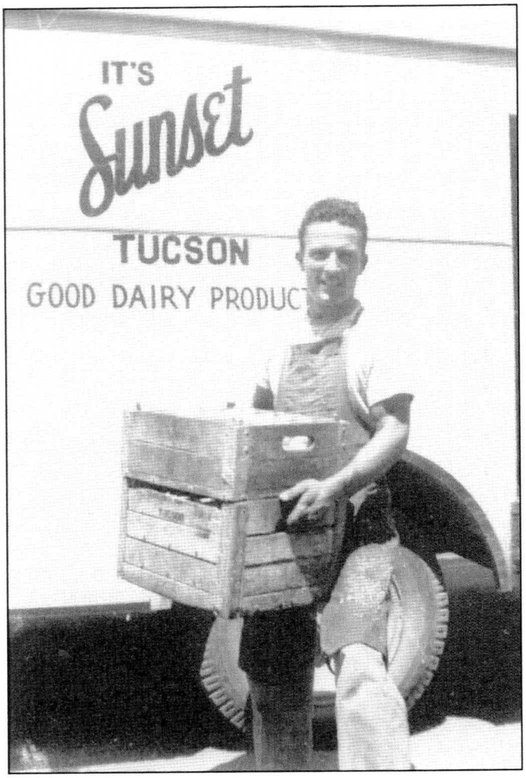

Max Mitchel delivers for Sunset Dairy in the 1960s. Mitchel worked for Shamrock Dairy and Sunset. Milk delivery would start at 1:00 a.m. to avoid the heat. Families often slept out in their yards because it was too hot in their houses. Deliverymen had to be careful not to step on any of the sleeping customers. Mitchell was the husband of Venla Gardner Mitchell of the Gardner family on Gardner Lane. Venla started out working at Homer Davis Elementary (which her kids attended) and retired as the school district's controller under John J. Bauman, who started in the district in 1959. (Loraine Mitchell Gerovec and Venla Mitchell.)

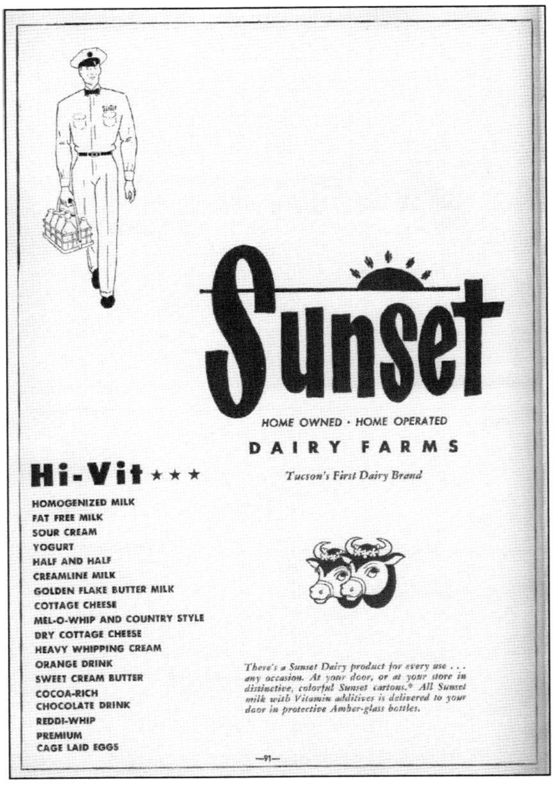

William A. Knapp moved to Tucson in 1915 and bought property by Jaynes Station on the Casa Grande Highway (now Interstate 10). This was the one and only home he owned. Adjacent to his home, he started his first diary, Sunset Dairy, in 1921. Knapp's son Bruce grew up in the dairy business and on the farm. He took over the dairy in 1944. He attended the original Laguna School (now in the Flowing Wells School District). Bruce ran Sunset Dairy until the late 1960s, when it closed. The property was sold in 1997. (FWUSD.)

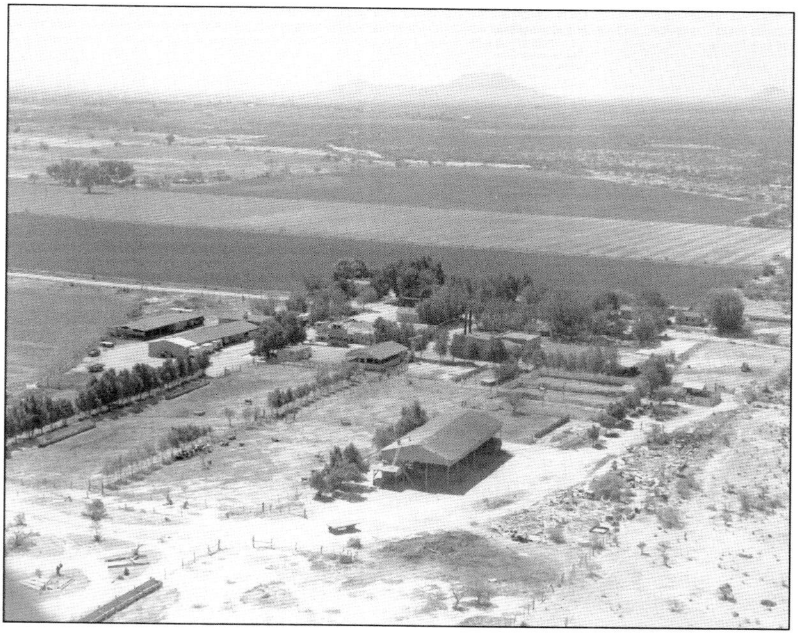

This c. 1950s aerial photograph shows the Sunset Dairy. The dairy is east of the Santa Cruz River. Sentinel Peak Mountain can be seen in the background to the south. The dairy property had numerous shade trees, which helped on the hot summer days. (Arizona Historical Society.)

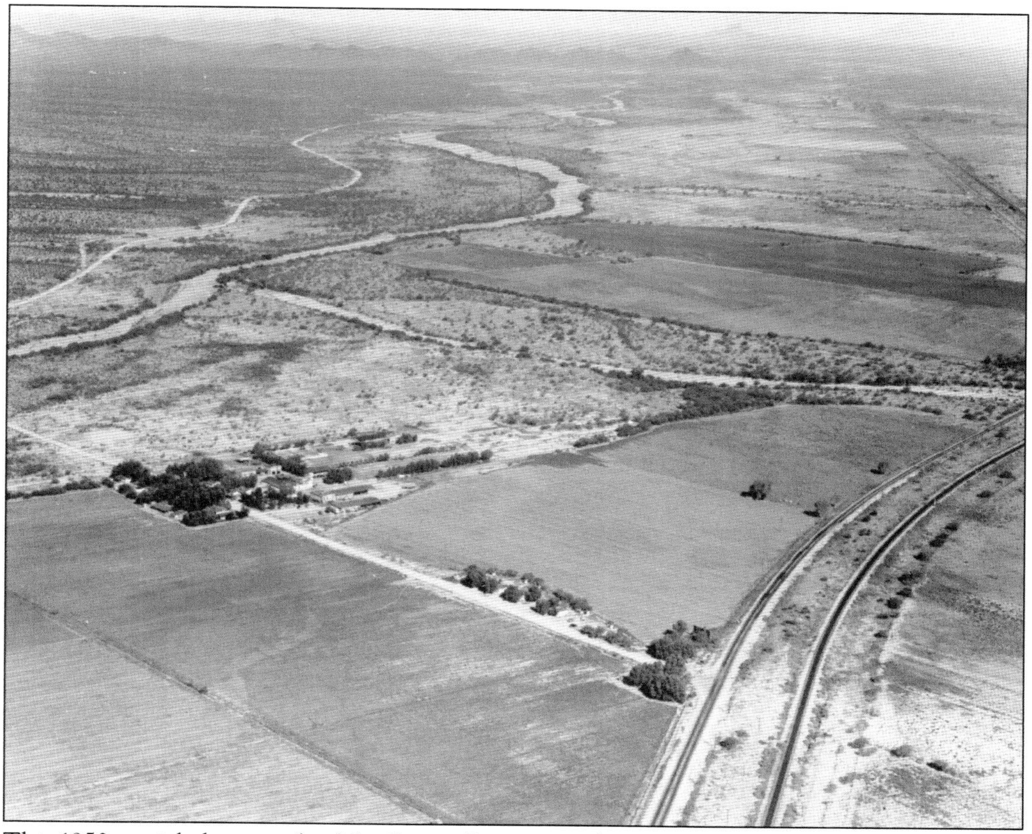

This 1950s aerial photograph of the Sunset Dairy was taken facing northwest. The Casa Grande Highway is at lower right. Sunset Road runs through the center. The Santa Cruz River is at upper left. (Arizona Historical Society.)

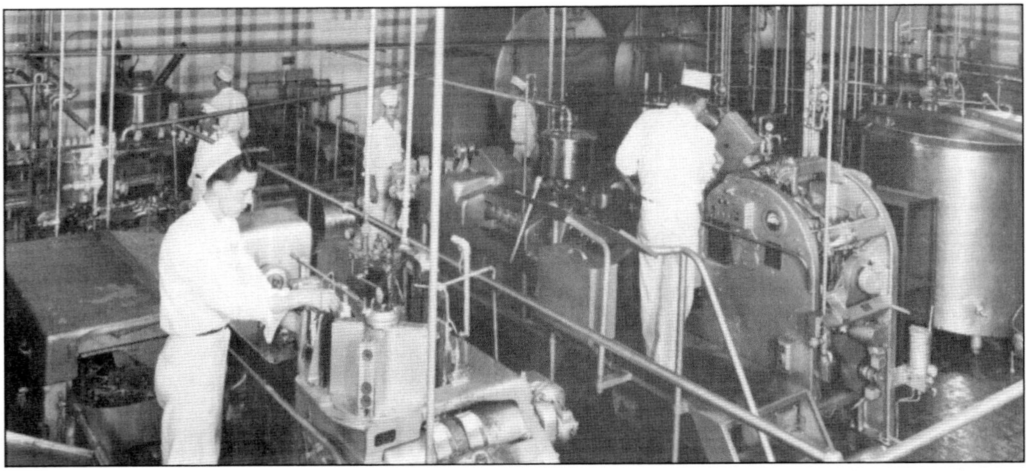

Shamrock's production controls assured the highest-quality products. Workers keep watch over the machines while processing the milk in 1954. Field trips to the dairy were a regular event at Flowing Wells School. The dairy was founded and owned by William T. McClelland and his wife, Winifred Parker McClelland. Winifred's brother William J. Parker was a general manager and later served as chairman of the board until 1982. A street at Flowing Wells Junior High is named for William. (FWUSD.)

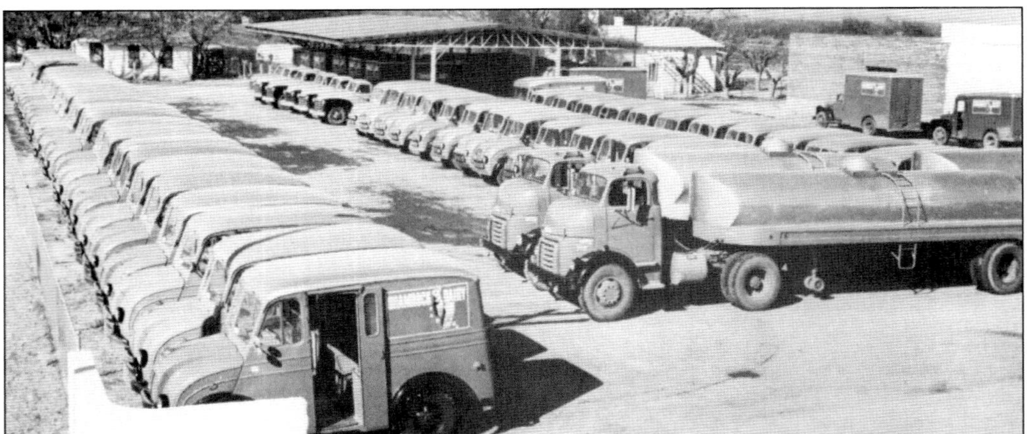

William T. McClelland (1891–1968) and his wife, Winifred Parker McClelland, bought 40 acres off Ruthrauff Road near Romero Road and started what was originally called the New Modern Dairy with a Model T truck and 20 cows. He sold the dairy in 1920 but regained control in 1922 due to lack of payment and renamed it Shamrock Dairy. W.T. ran the farm, while Winifred did the bookkeeping and sometimes drove the delivery trucks. In 1941, the dairy expanded to a new building. Over time, it added 300 acres and a fleet of GMC delivery trucks and Divco milk trucks (seen here) to deliver milk and other dairy products to families and stores. (FWUSD.)

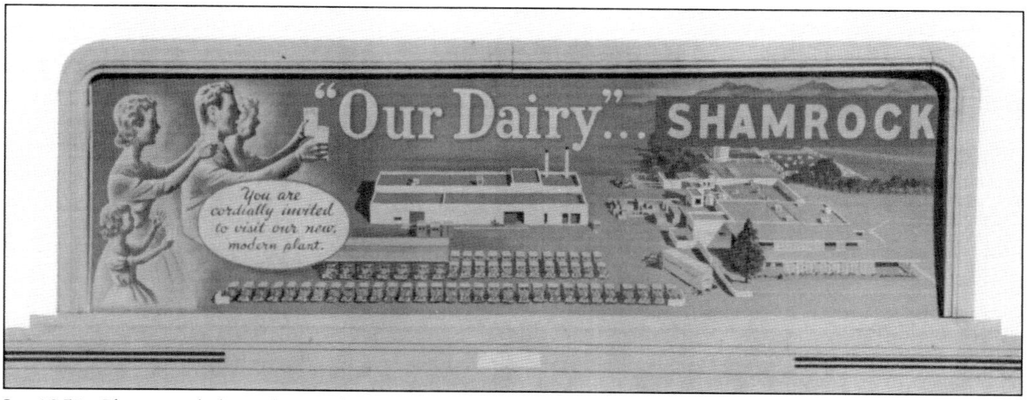

In 1951, Shamrock bought its first refrigerated truck. In 1949, the McClellands' son Norman (1927–2017) joined the business. He expanded the delivery area into Douglas and Bisbee. In 1956, Norman urged the building of Shamrock in Phoenix. Shamrock moved headquarters from Tucson to Phoenix after building a processing plant and renaming it Shamrock Foods. Shamrock was a major supporter of the Flowing Wells School District, and many employees' kids attended the school. Shamrock also contributed scoreboards at Flowing Wells. As supermarkets started appearing, milkmen and glass bottles were replaced by cartons and plastic bottles. The Ruthrauff Road location is still in operation today and is used as a distribution center. Norman remained chairman of the board of Shamrock Foods until the time of his passing. (FWUSD.)

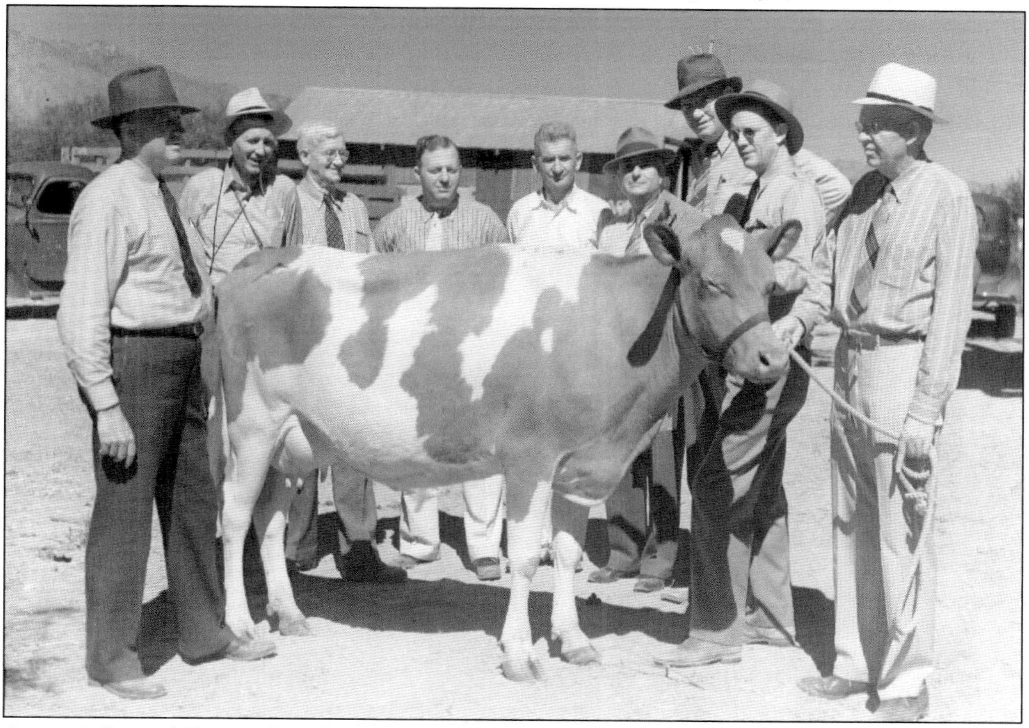

Arizona's Second Outstanding Guernsey Cattle Judging Meet was held at the Shamrock Dairy on May 10, 1947. From left to right are Prof. E.C. Gordon, University of California; Clyde Rowe, University of Arizona; Alec Wilson, manager of the Pomona, California, fair; C. McFarland, Tucson High School; W.T. McClelland, Shamrock Dairy; H.E. Geare, Phoenix; L.D. Kelmedson, state director of vocational agriculture; Jack Robinsen, Southwest District representative of the American Guernsey Cattle Club; and Prof. R.N. Davis, University of Arizona. (Arizona Historical Society.)

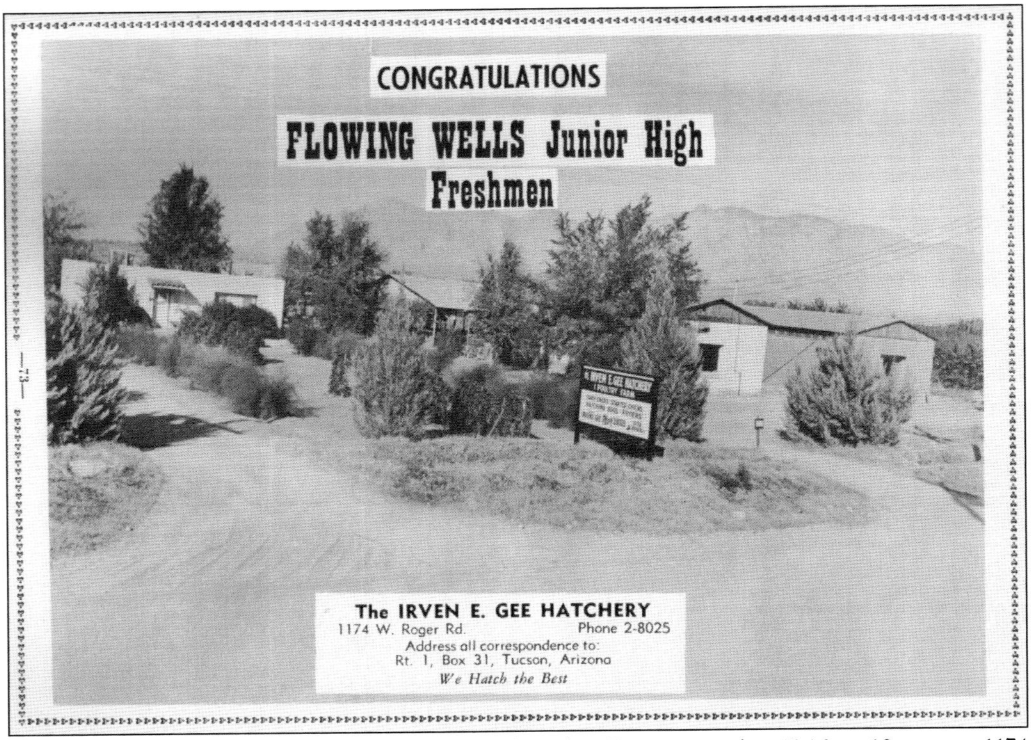

Pictured in 1954, the Irven E. Gee Hatchery and Poultry Farm opened in 1946 on 10 acres at 1174 West Roger Road (at Reno Avenue). The hatchery offered baby chicks, starter chicks, hatching eggs, and fryers. Jim Elliott remembered that since many of the neighborhood kids did not live on a farm, it was fun to help out at the Gee place as they planted alfalfa, which they harvested, bailed, and sold. They liked helping son Wilson Gee (FWHS 1957) with his farm chores, including setting irrigation siphon tubes and stacking bales on the tractor-pulled wagon. For them, it was fun. For Wilson, it was required. The hatchery closed in 1960. (Geraldine Gee Larrington.)

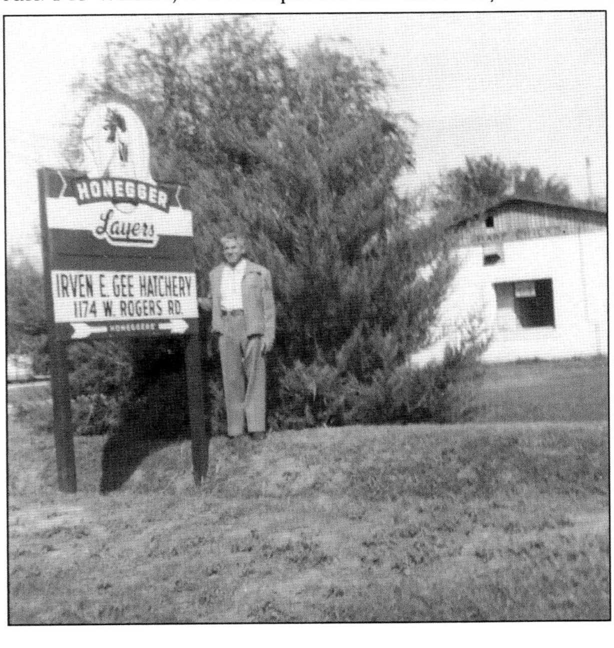

Irven Gee had a lifetime of experience and professional training in poultry. He graduated from the Poultry Department of the University of Arizona with distinction in 1932 and did graduate work the following year. For five years around 1950, he was the farm manager of the Poultry Feed Training Farm of the Arizona Flour Mills. (Geraldine Gee Larrington.)

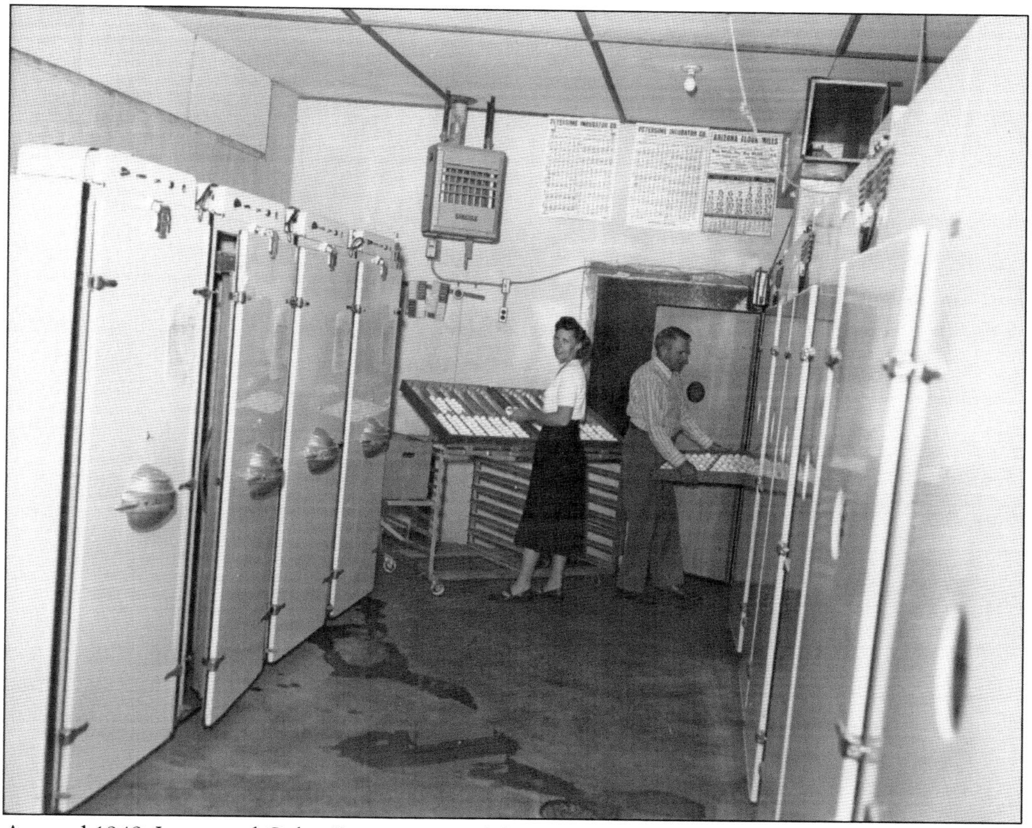

Around 1949, Irven and Celia Gee, owners of the Irven E. Gee Hatchery and Poultry Farm, tray eggs to be placed in the Jamesway incubators. Jamesway Manufacturing designed ultramodern incubators in which the eggs were carried through to completion of incubation without interruption in temperature, ventilation, or humidity. (Geraldine Gee Larrington.)

Around 1949, Irven and Celia Gee stand in front of the Jamesway incubators at the Irven E. Gee Hatchery and Poultry Farm. Irven and Celia met in 1930 and married in 1933. Irven served as president of the Arizona Poultry Association. He also served two three-year terms on the Flowing Wells School Board between 1948 and 1957. Celia was recognized with a lifetime membership in the Flowing Wells PTA for volunteering much of her time away from the hatchery to the school. Her daughter GeeGee said, "She sewed many of our clothes including yoked western shirts with cuffs and pearl snaps, squaw dresses and a memorable velvet coat." (Geraldine Gee Larrington.)

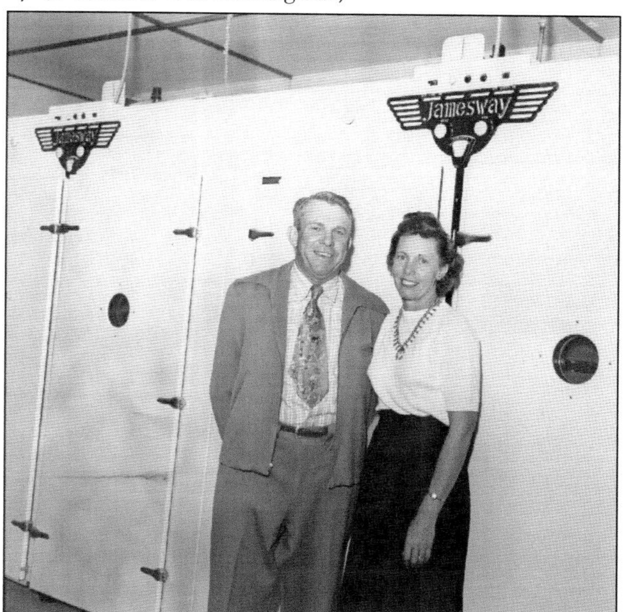

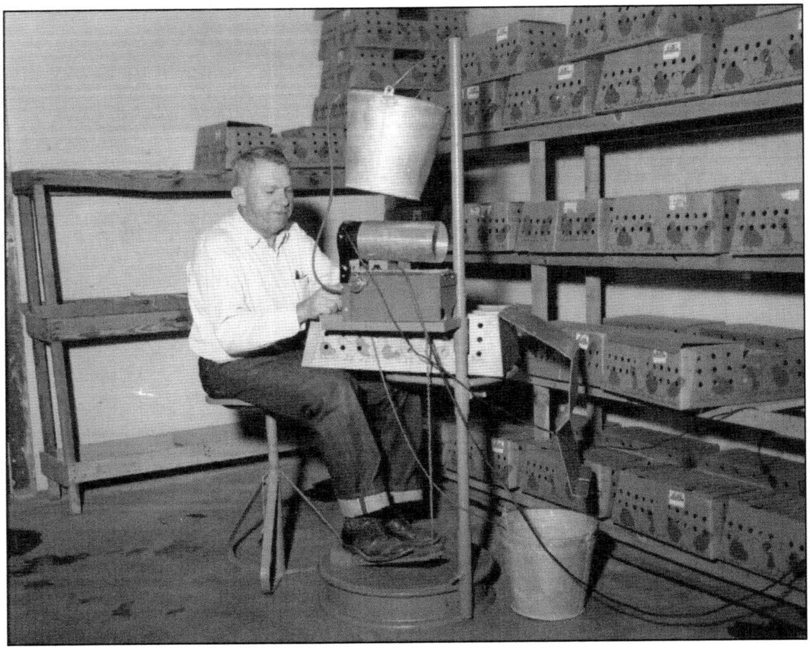

Irven Gee is shown operating a debeaker around 1949. This machine would clip off the tip of each chick's beak and cauterize it using a sharp, hot metal blade. Debeaking was done to prepare the chicks for shipping to customers. (Geraldine Gee Larrington.)

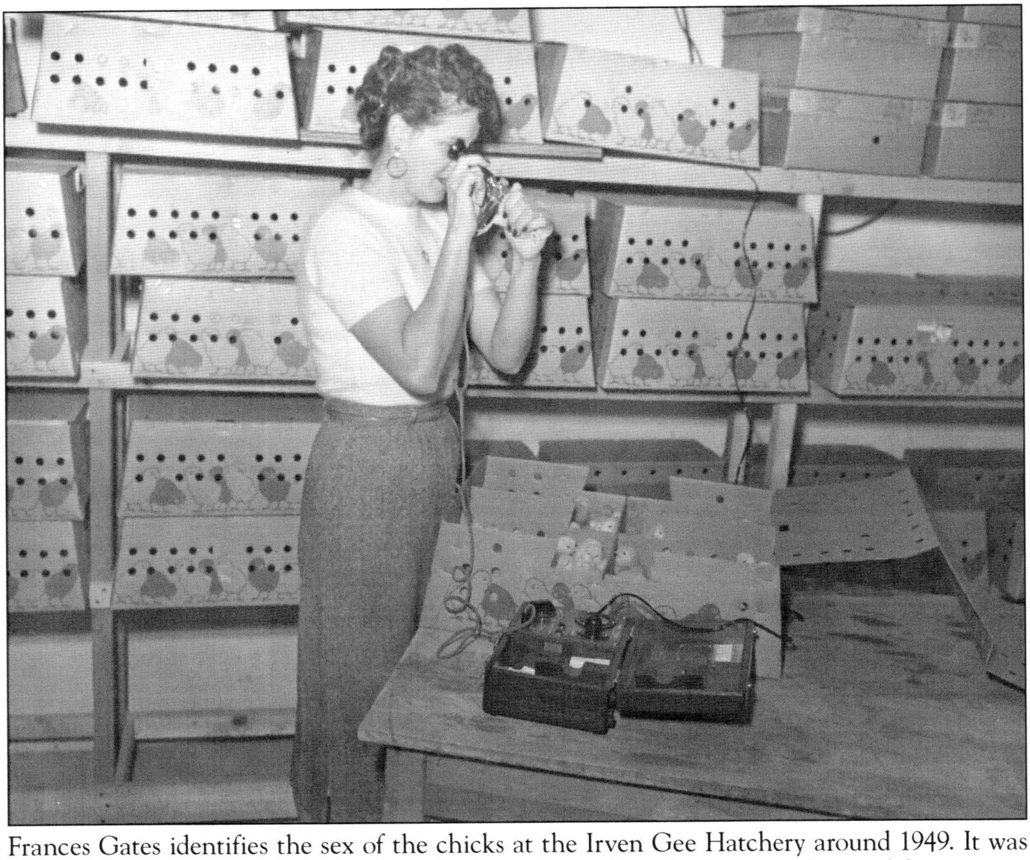

Frances Gates identifies the sex of the chicks at the Irven Gee Hatchery around 1949. It was important to determine the sex of the chicks, as laying hens had more value, and few roosters were needed to maintain fertility in the eggs for hatching. (Geraldine Gee Larrington.)

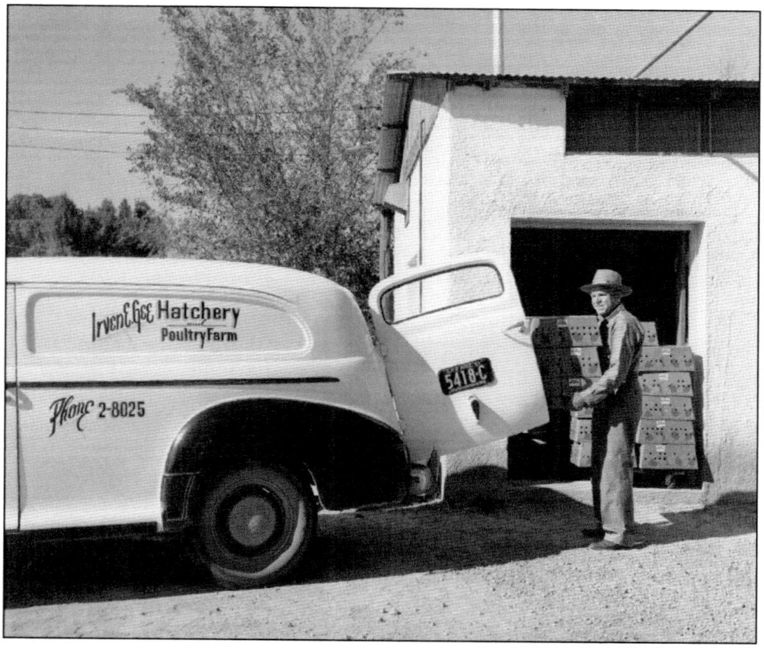

Irven Gee would often be seen loading chicks into a late-1940s Chevrolet panel delivery truck at the hatchery. Chicks were delivered to the Railway Express Agency (REA) at the train station on Toole Avenue. They were shipped to Mexico, around Arizona, and elsewhere. Julian Aguilu, who taught Spanish at Flowing Wells Schools, often translated letters to and from Mexico for Irven Gee. (Geraldine Gee Larrington.)

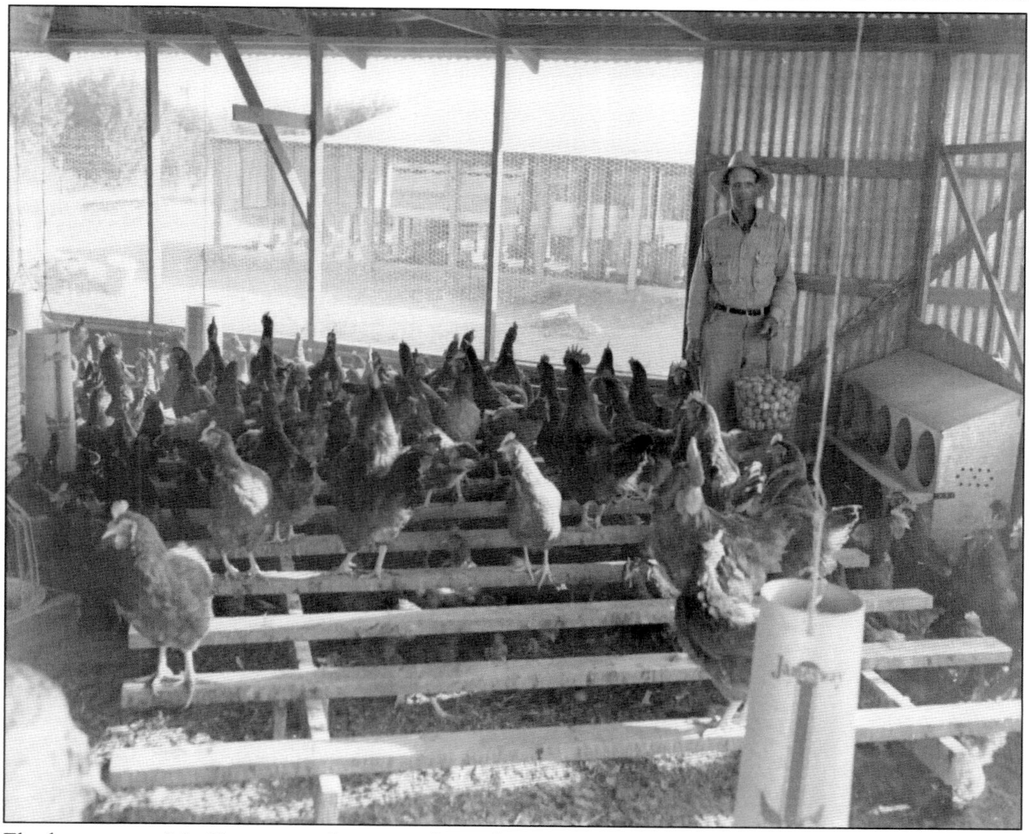

Flock manager Mr. Barnett gathers eggs from the New Hampshire breeding flock at the Irven G. Hatchery on Roger Road around 1953. Barnett took pride in his flock, which had an outstanding egg-laying record. (Geraldine Gee Larrington.)

$ $ SPECIAL $ SPECIAL $ SPECIAL $ $

Dear Neighbor:

This special introductory offer means dollars to you. This card will be honored only during February 1951 with a 10% discount on all chicks bought or ordered during this month at our regular competitive prices. Buy your chicks from us and avoid shipping losses.

We are the only Arizona hatchery licensed to produce the NORTHWESTER chick, the new, sensational, hybrid that combines all the best qualities of its immediate ancestry; delicious pheasant and the best strains of meat chickens. The Tucson Poultry Processors are offering a nice premium on NORTHWESTERS over all other breeds. Ask them about the NORTHWESTER. Also, we have the finest all-purpose New Hampshire chicks. Price list on request.

The Irven E. Gee Hatchery and Poultry Farm

1174 W. Roger Rd., Tucson Phone 2-8025

In 1951, the Irven E. Gee Hatchery and Poultry Farm offered an introductory special to customers who ordered before the end of February. (Geraldine Gee Larrington.)

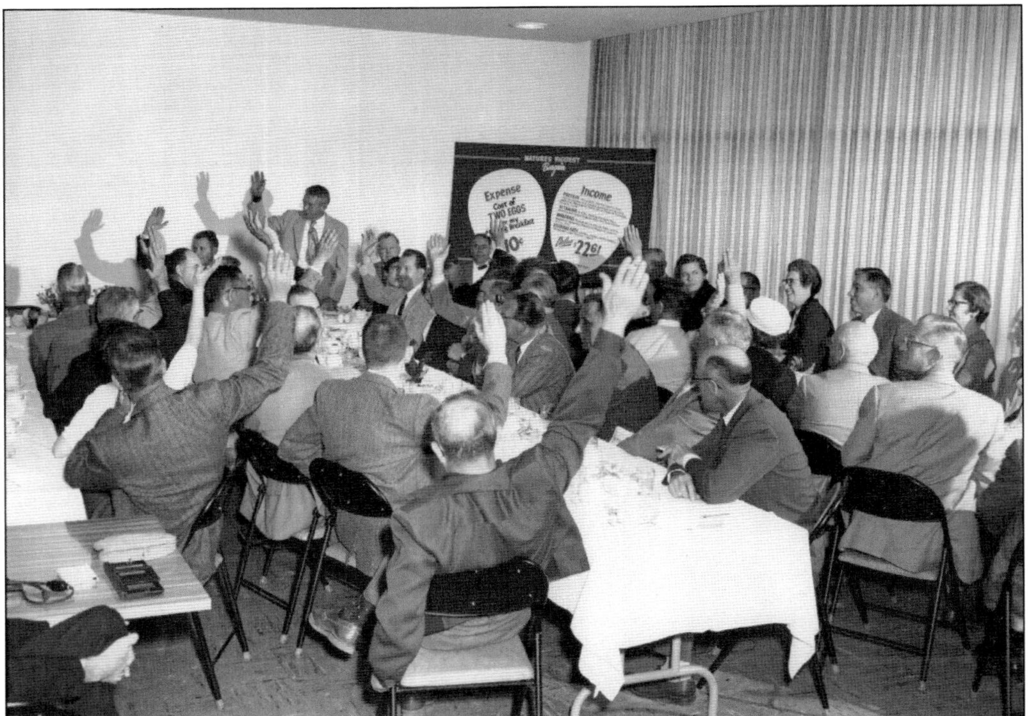

Members of the Arizona Poultry Association hold a meeting in 1953 at the Shamrock Dairy conference room. Irven Gee served as president. His hatchery, along with Shamrock Dairy and Arizona Feeds, united in a cooperative effort to discuss industry issues. The Arizona Poultry Association is believed to have been established in the late 1920s to serve the organizational and educational needs of poultry farmers in the state. (Geraldine Gee Larrington.)

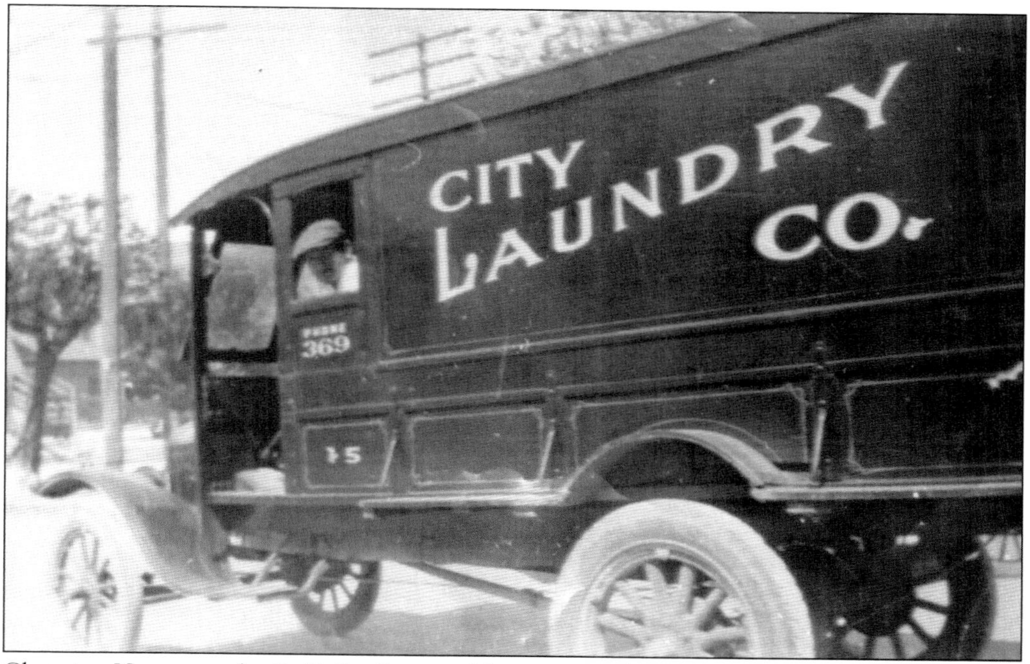

Christian Hansen and wife Nellie Eugenia Hansen moved into the Flowing Wells area around 1914 from Minnesota, drawn by the Tucson Farms Company advertising farming in the Santa Cruz Valley. The Hansens built their home on the northeast corner of Flowing Wells and Roger Roads, and it still stands there today. When the water situation was not as promised, Christian Hansen worked as a florist and for the City Laundry and Dry Cleaning Company downtown. Later, when water was sufficient with the Flowing Wells Irrigation District, the Hansens operated a family dairy. The City Laundry and this 1920s one-ton Ford Model T delivery truck were owned by James Plunkett. (FWUSD and Randy Holladay.)

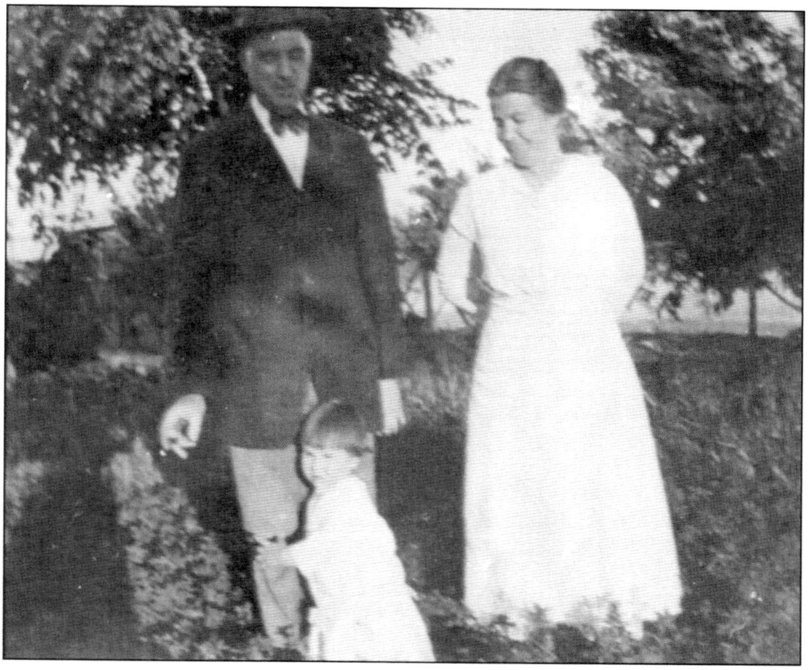

In 1908, banker George Bodell moved his family to the area. Bodell worked for the Valley National Bank from 1909 to 1955. He was involved in providing Valley Bank loans to many local residents and businesses. (FWUSD and Randy Holladay.)

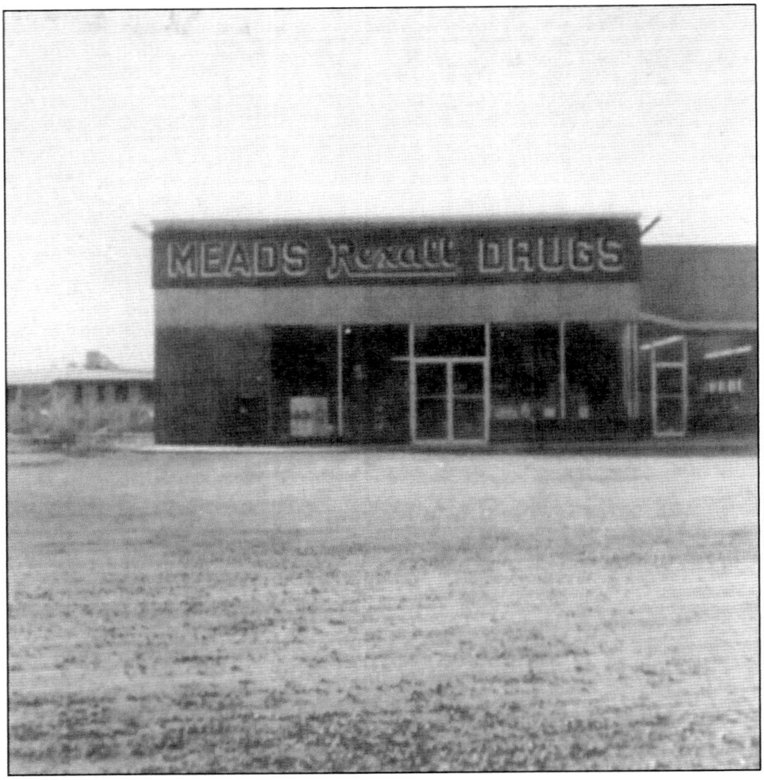

John M. Mead opened the original Mead's Rexall Drugs in 1957 at 1513 West Prince Road, across from the Iola Frans School. He decided to open a pharmacy of his own because "I worked in my uncle's pharmacy in Connecticut and I liked it. So I found a building," said Mead. He remembered going to the Iola Frans School during the polio scare of the 1950s to give children a sugar cube with the Salk vaccine in it. (FWUSD.)

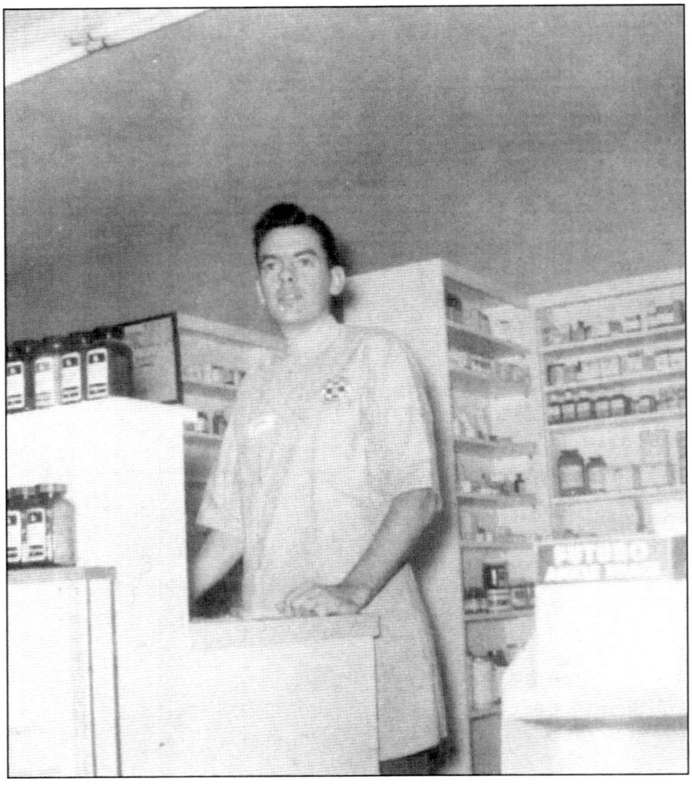

Pictured in 1958, John Mead moved to Tucson in 1945 due to his mother's health. He attended the Pharmacy College at the University of Arizona as one of the first four-year graduates in the program in 1951. He opened his first pharmacy on West Prince Road. In 1961, he moved Mead's Pharmacy to the corner of Prince and Flowing Wells Roads. John said he "filled 300 prescriptions a day using a typewriter for the labels." Mead served on the Flowing Wells School Board from 1963 to 1977. He filled his last prescription in 1997, when he closed the pharmacy after 40 years in business. Mead was inducted into the Flowing Wells Athletic Hall of Fame in 2006. (FWUSD.)

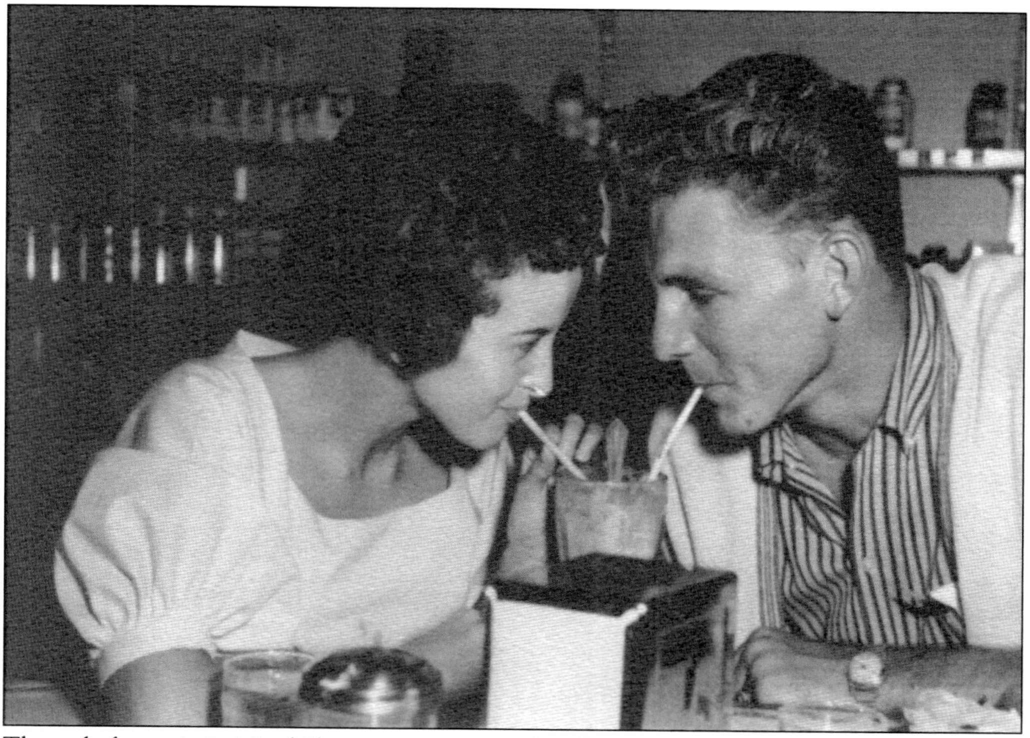

The soda fountain in Mead Pharmacy was a popular place for kids to hang out. Brenda Slawson (FWHS 1961) and Bill Knight (FWHS 1960) share a soda around 1960. Mead added the soda fountain in the late 1950s to increase traffic in the store. His store also had the largest Hallmark card shop in Tucson. (FWUSD.)

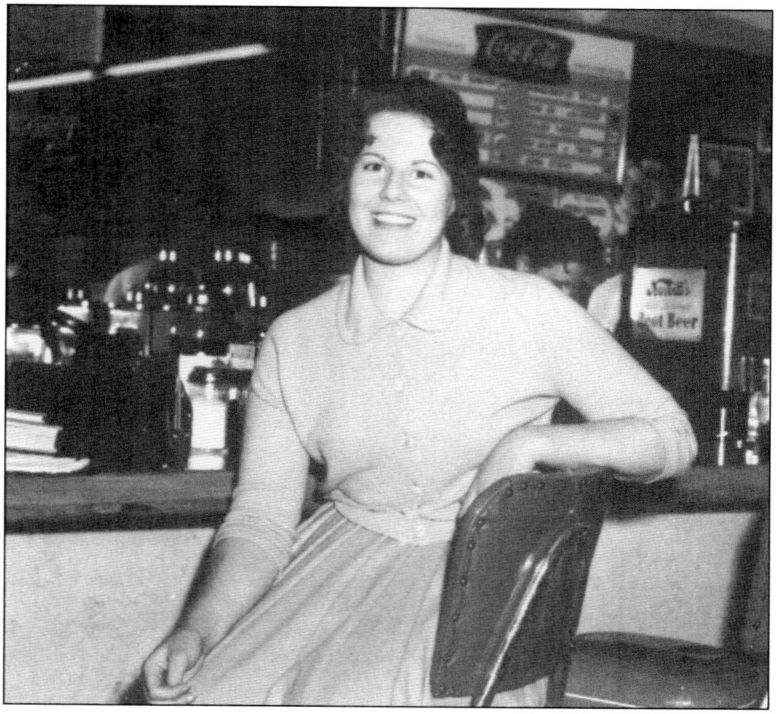

Cheryll Hitchman (FWHS 1962) spent time at the soda fountain at Mead's Pharmacy in 1961. "Mead's cheeseburgers were the best in the world," proclaimed Melanie Turner Black (FWHS 1964). (FWUSD.)

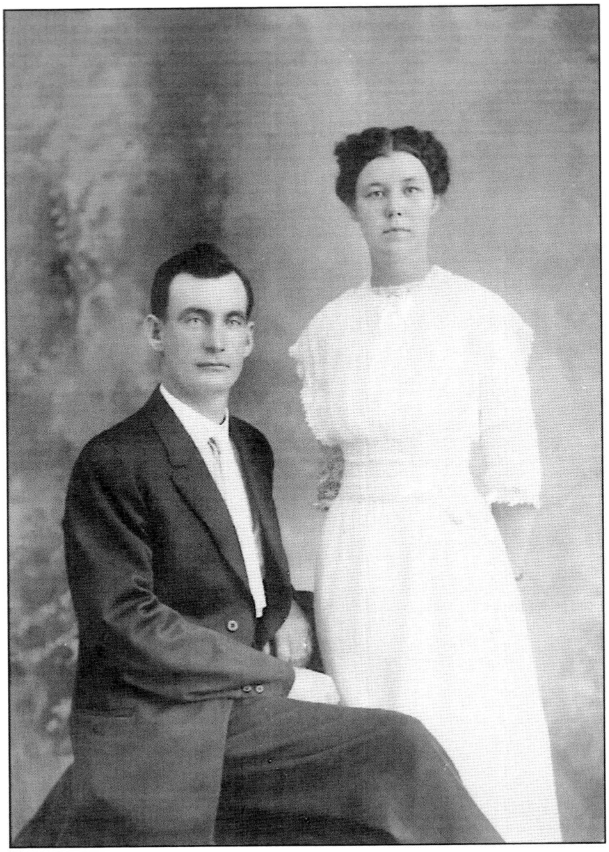

John Houston and Mary Estella Emberson Lawson's farm was on the modern-day FWHS baseball, practice, and softball fields. This land was on the south end of their farm. Their home and barn were on the north end of the property. John and Mary and their children moved to Tucson from Beaver County, Oklahoma. It is rumored that those who were old enough to drive each drove a Ford. John opened a Ford dealership with those cars. The children, from oldest to youngest, were Charles Edwin, Leonard Cecil, Irene Elizabeth, John Eldon, Mary Evelyn, Ina Gertrude "Trudy," Myrtle Estella, Georgia Lou, Glenn, Dale, and William "Bill" Houston. Several of them were teachers in Tucson. Glenn Lawson taught music at Flowing Wells. (John Lawson.)

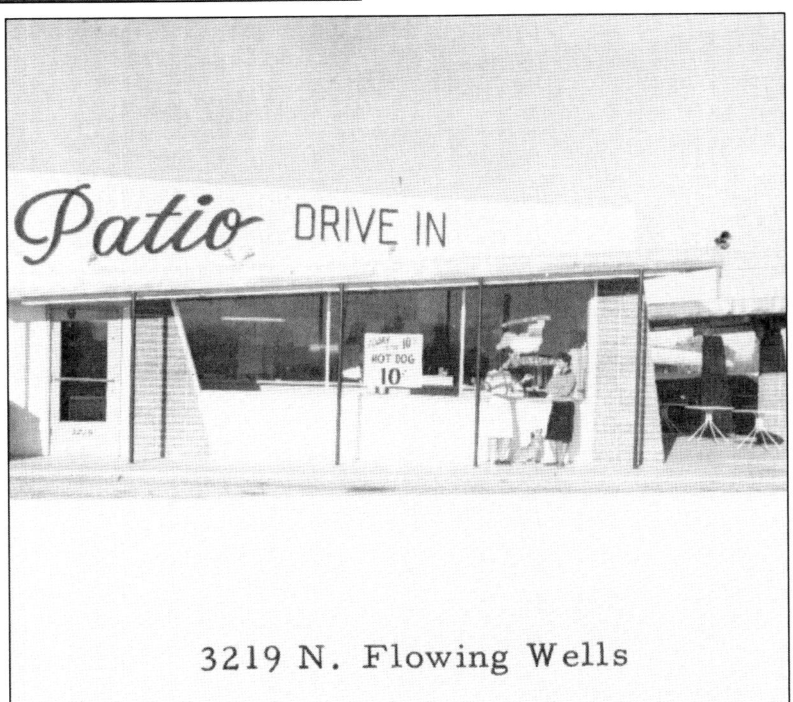

The Patio Drive In was built around 1961 and owned by Howard and Mary Denman. The Patio was known as the "Home of the Monster Burger" in the 1950s. Hot dogs were 10¢ on the 10th of the month around 1961. The Patio was at the corner of Flowing Wells Road and Sunshine Lane (now West Fort Lowell Road). (FWUSD.)

Jack Ellis Sporting Goods opened in 1951 and is pictured around 1960. Jack sold his interest in the store to his brother Charles in 1953. Charles passed away in 1969. Charles's wife, Jean Ellis, sold the store in 1974 to the owners of Speedway Sports Shop. Because the Jack Ellis name was well known in southern Arizona and had a good reputation, the new owners kept the name and expanded to six stores. The store changed ownership again in 1982 with four of the six locations in operation. Jack Ellis Sporting Goods closed in 1984 due to financial stress and competition from national chain stores. It was one of two stores where locals, including the school, purchased sporting goods. Another store, Porter's, sold uniforms to the district. (FWUSD.)

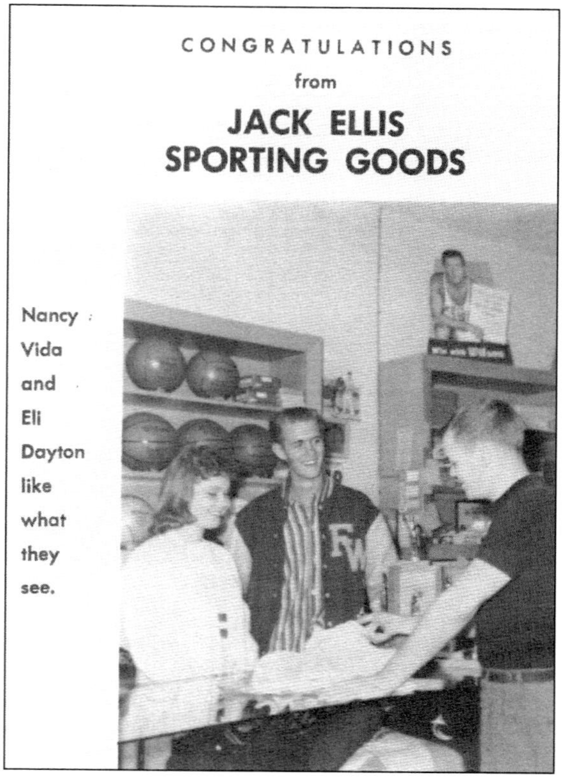

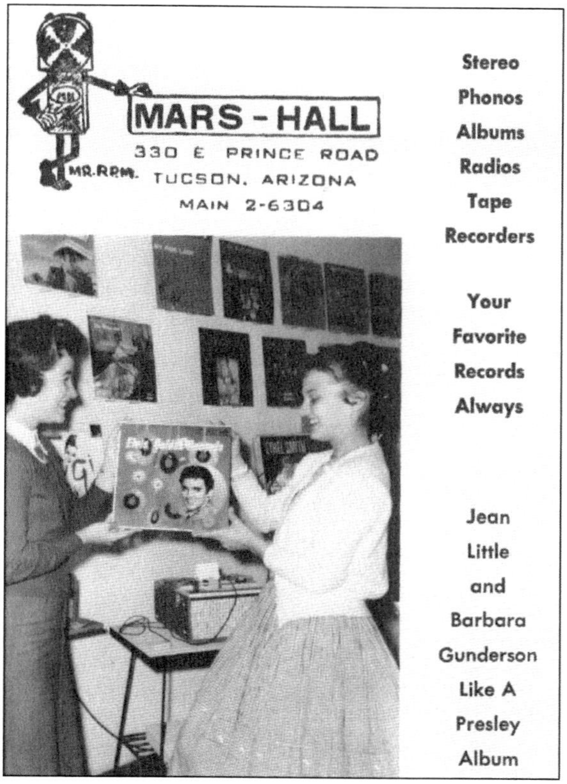

Mars-Hall was opened for business in 1954 by Athene Poulos and her husband. Her goal of pleasing the customer kept her in business for over 44 years. Many of her customers were students. It was reported that in 1955 Mars-Hall sold fifty 78-rpm copies of "Rock Around the Clock" by Bill Haley and the Comets in 30 minutes. Here, FWHS students Jean Little (left, FWHS 1962) and Barbara Gunderson (FWHS 1960) shop for Elvis Presley albums around 1960. (FWUSD.)

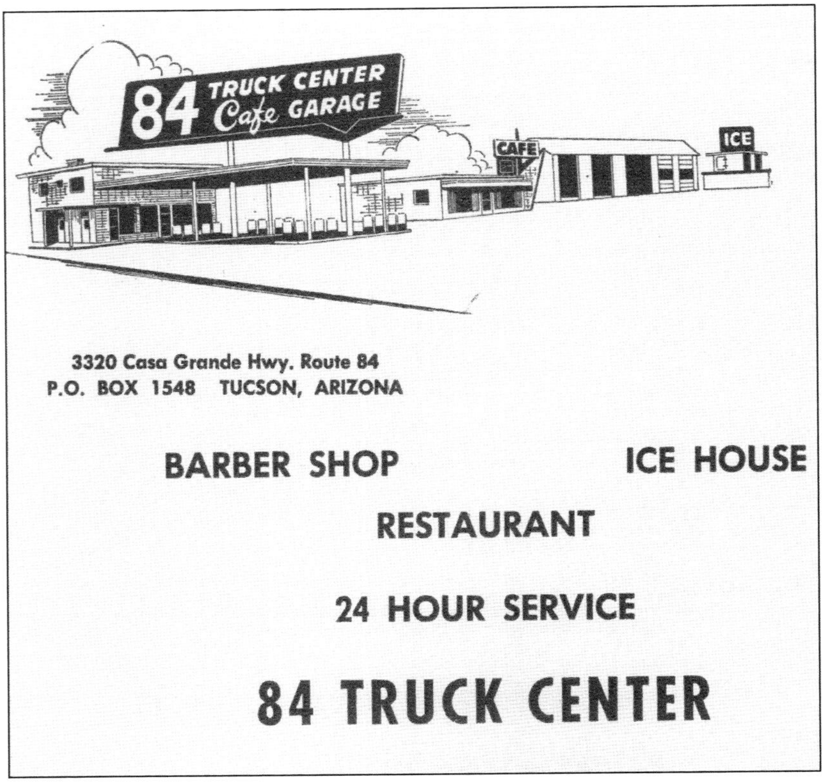

In 1959, there was not much along Highway 84 except for the 84 Truck Stop at 3320 North Casa Grande Highway south of Prince Road. Men could stop and rest in its 20-bed, second-floor dormitory and eat and shower in an air-conditioned facility. Locals enjoyed pie at the 24-hour café after a football game. (FWUSD.)

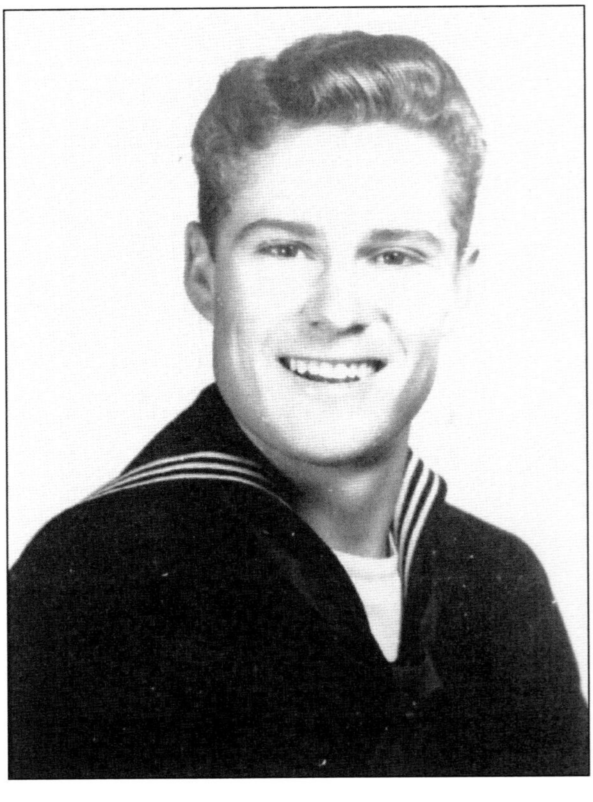

Young's Welding, at 1050 West Prince Road, was originally opened in 1951 by Ira Smith. Known to locals as "Smitty," he bought 13 acres at the site for $6,500. Burl Young moved to Tucson in 1936 at the age of six; he started welding at 17. After serving in the Navy and the Reserves from 1948 to 1953, he started working for Smitty in 1955. Seven months later, he bought the shop. "In the 1950s people were starting to pull camp trailers, so I specialized in trailer hitches," Young said. When he bought the shop, there were open irrigation ditches on both sides of Prince Road to supply water to the farms. He is still welding there today. (Burl Young.)

Food Giant Market opened in 1951 on Prince Road. It was the first supermarket in the area. The store gave out S&H Green Stamps with every purchase, with double stamps on Tuesdays. S&H stamps were popular from the 1930s until the late 1980s. The Sperry and Hutchinson Company, founded by Thomas Sperry and Shelley Byron Hutchinson, started selling stamps to merchants in 1896 with the idea that the stamps would build customer loyalty. Customers could trade in their stamps for products. (FWUSD.)

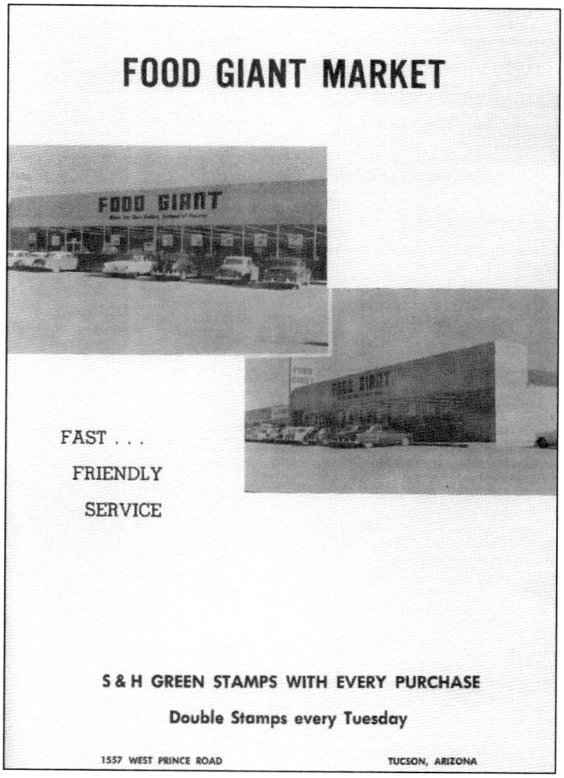

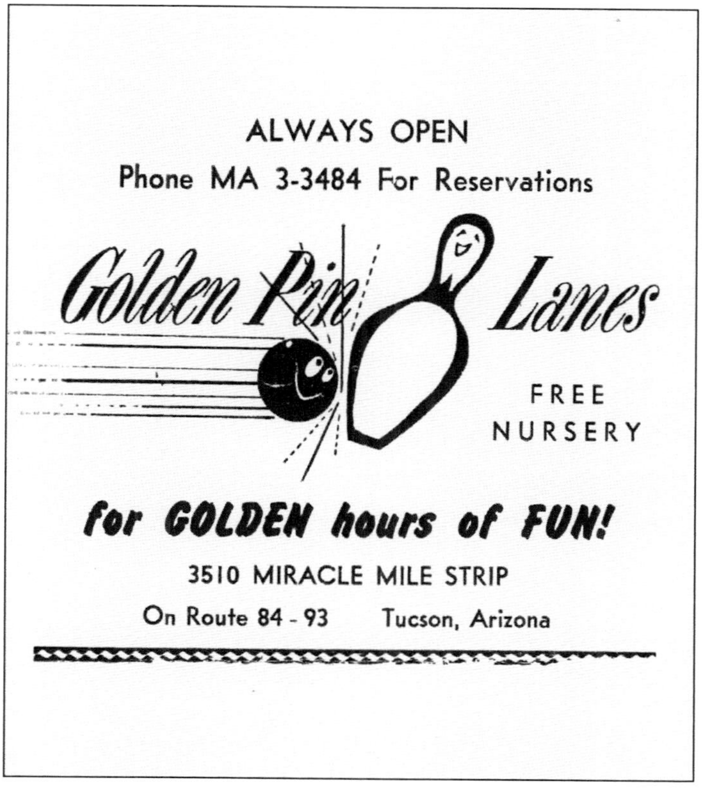

Don Allen started construction on Golden Pins Lanes in 1959 and opened for business in 1960 with 32 lanes on Miracle Mile. In 1976, an additional 16 lanes were added, making Golden Pins the largest bowling center in Tucson. Pete Tountas, three-time Pro Bowling Association (PBA) champion in 1968, managed the bowling alley for many years. Tountas was elected to the American Bowling Congress Hall of Fame. Golden Pins hosted the PBA Tour for more than 30 years. (FWUSD.)

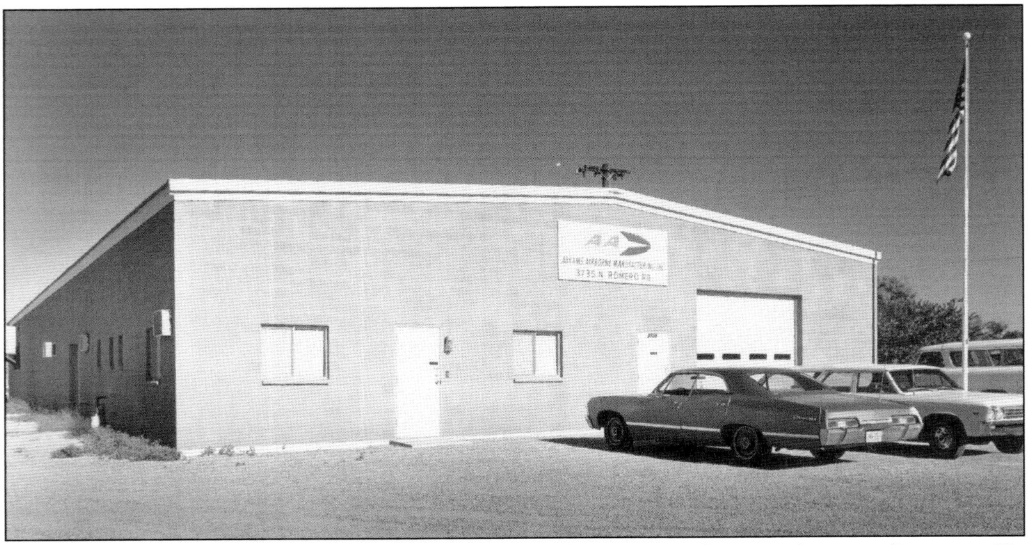

Harold "Bud" Abrams (Tucson High School 1941) always wanted to own his own business. He built an aviation empire from meager beginnings, a $50,000 loan, and a passion for flying. Abrams Airborne Manufacturing opened on April 1, 1965. Abrams occupies property that was part of the old Freeway Airport in the Romero–Prince Road area. The company specializes in precision sheet metal fabrication, aluminum dip brazing, CNC machining, and aerospace assemblies. "He was a self-taught mechanical genius. He worked it up into a family project," said Gary Abrams, chief executive officer of Abrams Airborne Manufacturing. All of the Abrams family worked there. Today, third-generation family members work in the business. (Gary Abrams, Barbara Abrams, and Jenny Abrams Wilson.)

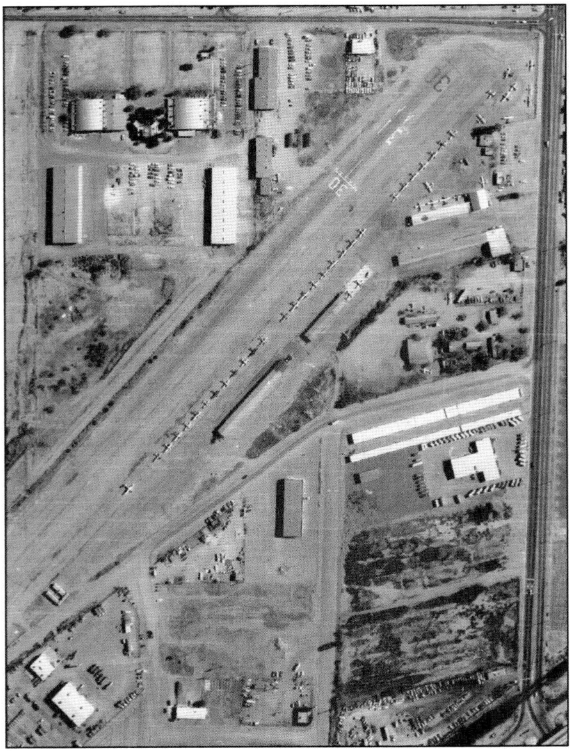

Freeway Airport is pictured in a 1973 aerial photograph. Prince Road runs vertically at right, and Romero Road runs horizontally at top. The Gilpin hangars and tower (now R.E. Darling Co.) are at upper right. Freeway Airport closed in 1978, when redevelopment of the property began, including an expansion of Abrams Airborne Manufacturing. In 1980, the end of the movie *Stir Crazy* was filmed at the closed Freeway Airport site near some of the still-standing hangars near Prince Road. (Gary Abrams, Barbara Abrams, and Jenny Abrams Wilson.)

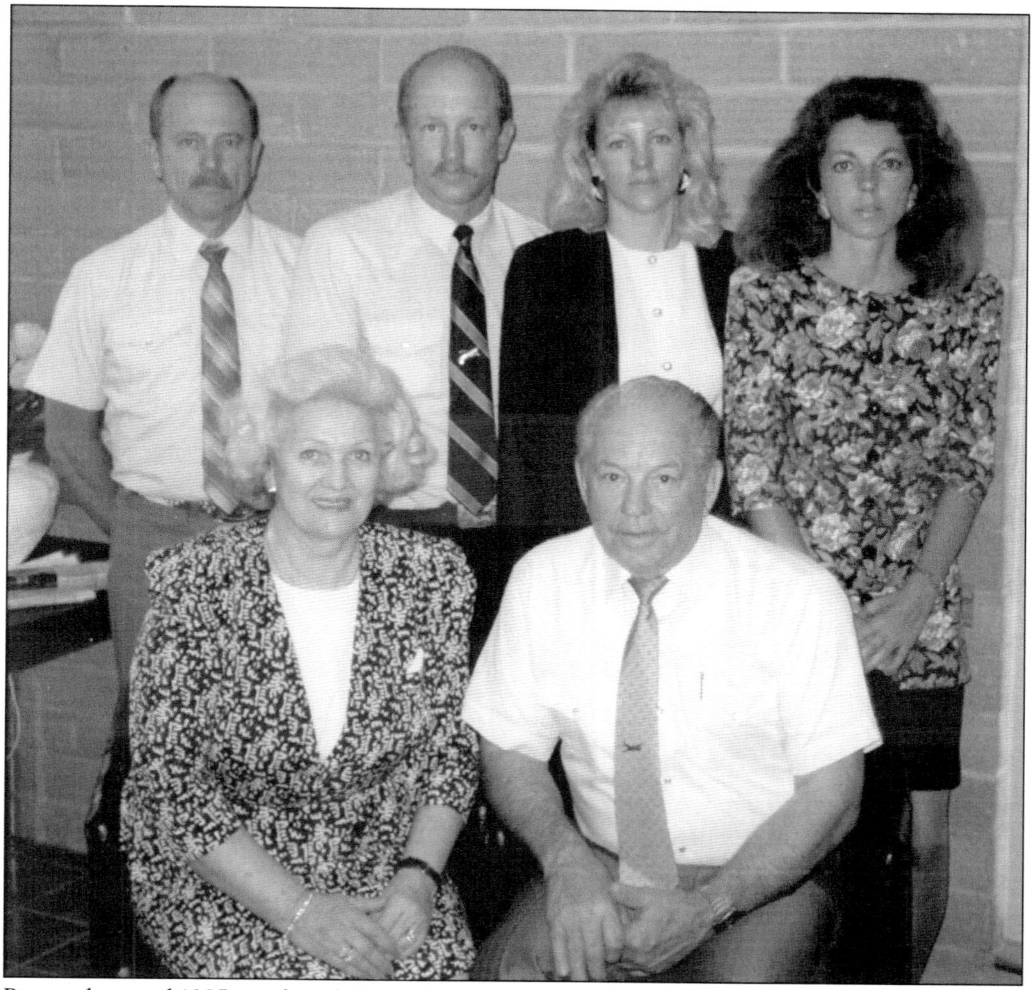

Pictured around 1995 are, from left to right, (first row) Barbara Abrams and Bud Abrams; (second row) Chris Abrams, Gary Abrams, Jenny Abrams Wilson, and Julie Abrams Sheppard. Harold "Bud" Abrams and his wife moved to Flowing Wells in 1948. Bud and Barbara, his wife of 59 years, grew up in Tucson and graduated from Tucson High School. Bud served in the US Army Air Corps as a pilot flying C-46s in the Pacific theater in World War II. Jenny Abrams Wilson said, "they decided to stay in the Flowing Wells area so their three children, Gary [FWHS 1970], Jenny [FWHS 1972], and Julie [FWHS 1974], could take advantage of the superior education the school district offered." The children began their education at Homer Davis Elementary School and graduated from FWHS. In 1955, Bud built a house on property at Wetmore and Romero Roads that had a Flowing Wells irrigation ditch running through it. The property was later sold to Pima County, and it currently houses the Ellie Towne Flowing Wells Community Center. (Gary Abrams, Barbara Abrams, and Jenny Abrams Wilson.)

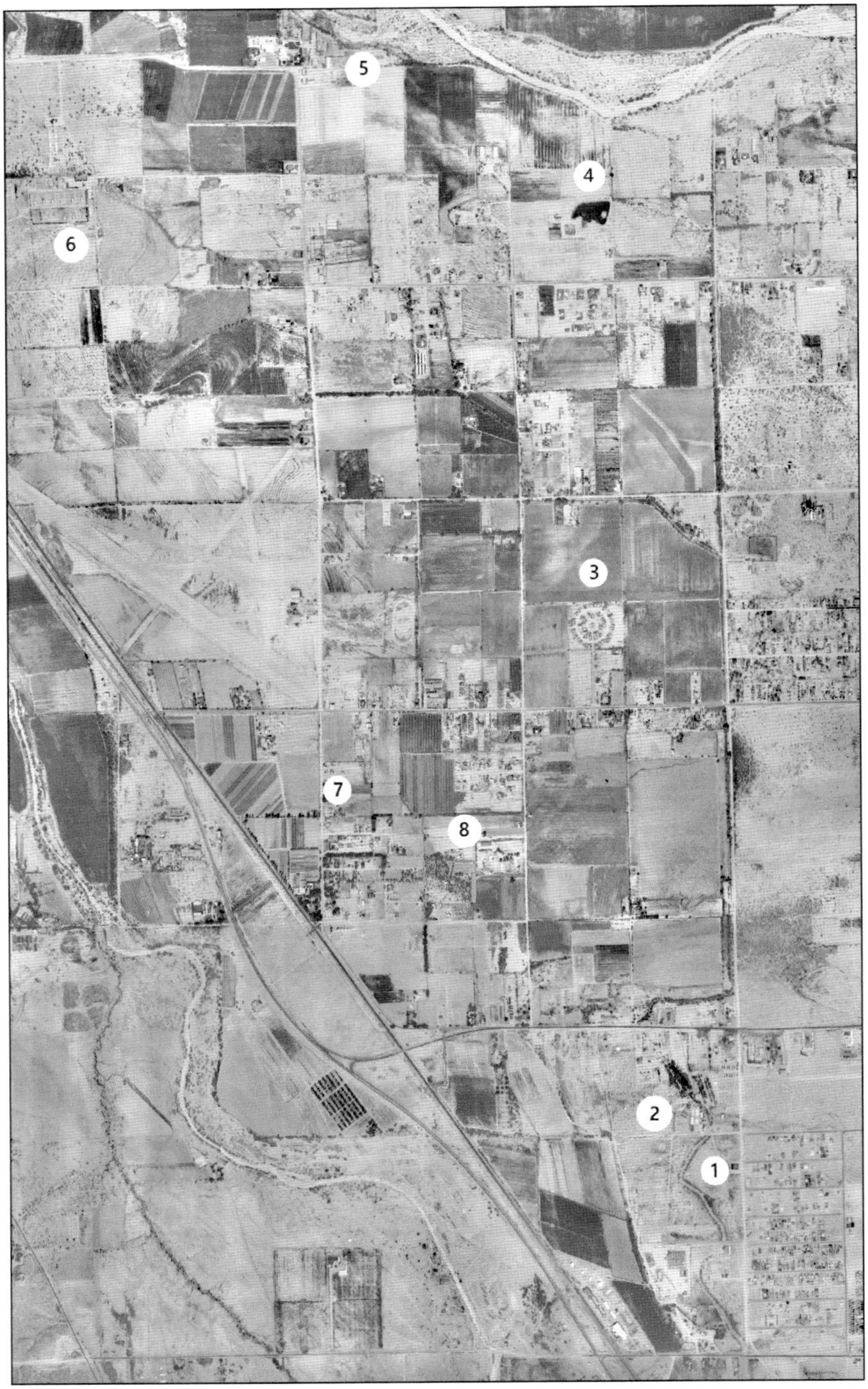

Pictured is a detailed 1953 aerial view of Flowing Wells from Grant Road (bottom) up to the Rillito River in the north. Several significant landmarks can be seen: 1. Allison Canal, 2. Desert Shores Lake, 3. Crescent Manor, 4. Lake Ranch (now Swan Lake) owned in the 1950s by Panama's vice-consul at Tucson, Leonardo Villanueva Meyer, 5. Shamrock Dairy, 6. Current site of Flowing Wells Junior High, 7. Basement outline remains of 1915–1928 Flowing Wells schoolhouse, 8. Site of Flowing Wells Ranch / Transient Camp c. 1933. (Cooper Aerial Survey and authors.)

Six

Flowing Wells Schools and Families

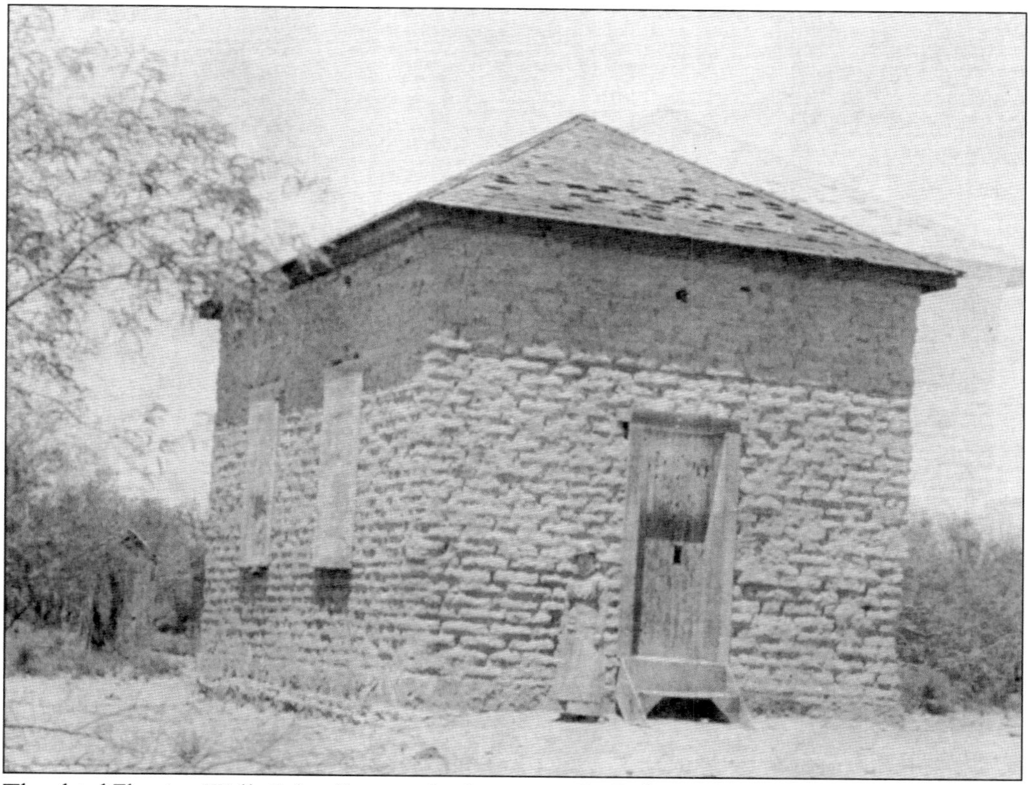

The third Flowing Wells Schoolhouse, also known as the Rillito School, is pictured in 1914 with Nellie Olsen. The original adobe building still stands on West Fort Lowell Road (Sunshine Lane) and Romero Road. The first and second schoolhouses were rooms in private homes. The first schoolhouse was in an adobe building in the center of the cotton farm around what is now Gardner Lane and the railroad tracks. The heads of the Harris and Bouschet families formed the school board. Mamie Bouschet married Henry Peterson (of the Holstein Dairy and a two-term Flowing Wells School Board member) in 1921. The first year (around 1968), the school had to be closed due to a smallpox outbreak, but it reopened in another private home the next year. Mamie Bouschet Peterson was one of the eight children in the first class. The school moved to the pictured location around 1900 and remained there until 1.5 acres were acquired from the Tucson Farms Company to build the redbrick Romero Road schoolhouse around 1915. Since all records were destroyed in the fire of 1938, Mamie provided oral history and does not refer to the time between the founding of the district and the first class she attended around 1896. (FWUSD and *Arizona Daily Star.*)

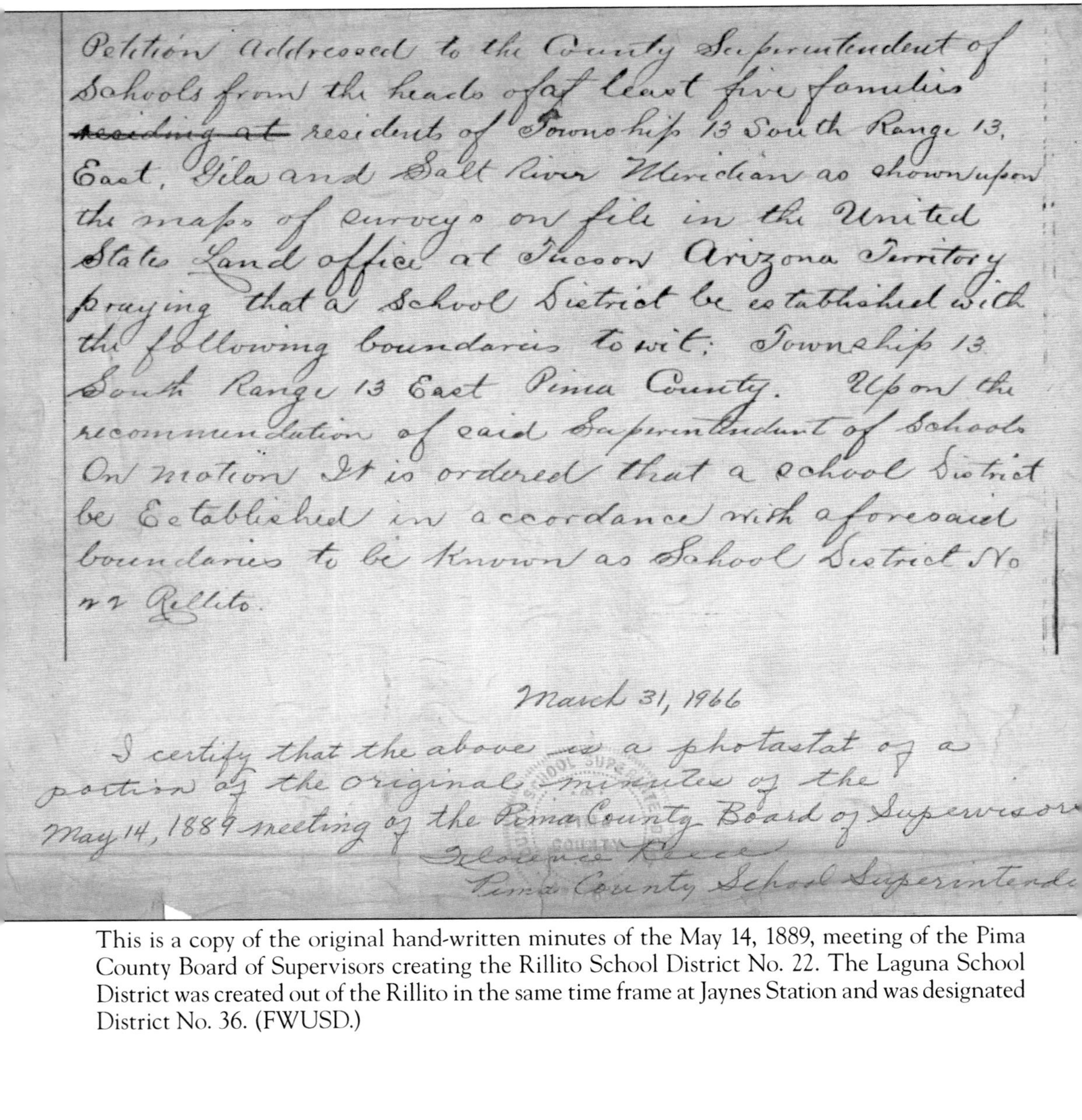

Petition addressed to the County Superintendent of Schools from the heads of at least five families ~~residing at~~ residents of Township 13 South Range 13, East, Gila and Salt River Meridian as shown upon the maps of surveys on file in the United States Land office at Tucson Arizona Territory praying that a School District be established with the following boundaries to wit: Township 13 South Range 13 East Pima County. Upon the recommendation of said Superintendent of Schools On motion It is ordered that a school District be Established in accordance with aforesaid boundaries to be Known as School District No 22 Rillito.

March 31, 1966

I certify that the above is a photostat of a portion of the original minutes of the May 14, 1889 meeting of the Pima County Board of Supervisors

Florence Reece

Pima County School Superintende

This is a copy of the original hand-written minutes of the May 14, 1889, meeting of the Pima County Board of Supervisors creating the Rillito School District No. 22. The Laguna School District was created out of the Rillito in the same time frame at Jaynes Station and was designated District No. 36. (FWUSD.)

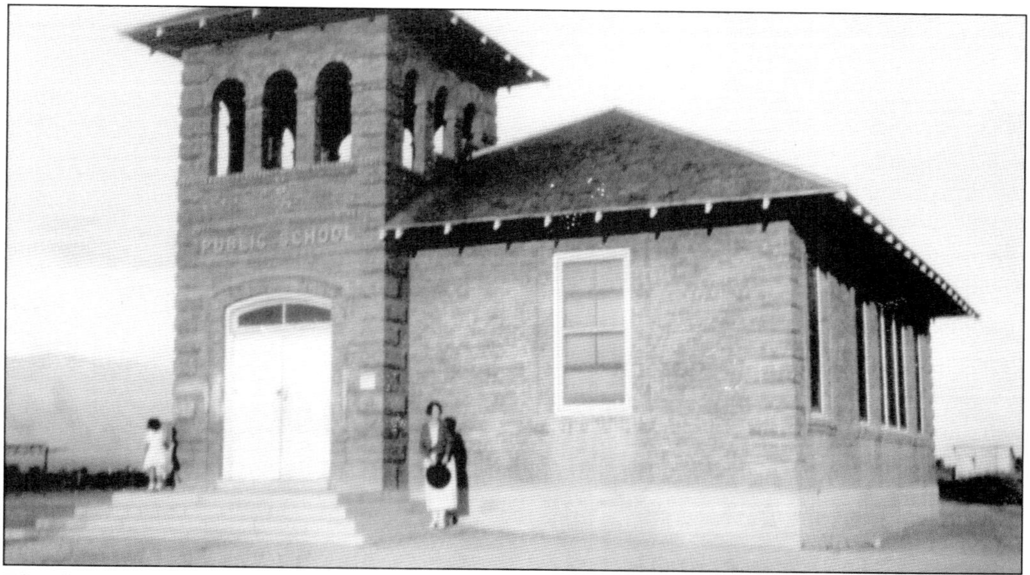

This fourth Flowing Wells Schoolhouse was built around 1915 with a $3,000 bond at 3404 North Romero Road on 1.5 acres bought from Tucson Farms Company. In 1928, two rooms, a basement, and a furnace were added. The building burned down in the early morning on Friday, October 28, 1938. The estimated loss to the district was $12,000. The school had three teachers: principal Edwin Van Doren, Ruth Downs, and Nellie Copeland. The fire left only the brick walls. The school had 70 students. The trustees decided to not join Amphitheater and moved to the recreation hall of the abandoned transient camp at 3300 North Flowing Wells Road, which was converted into classrooms. Students only missed one day of school while staff and residents worked for three days over the weekend. (Arizona Memory Project, Randy Holladay, and *Arizona Daily Star.*)

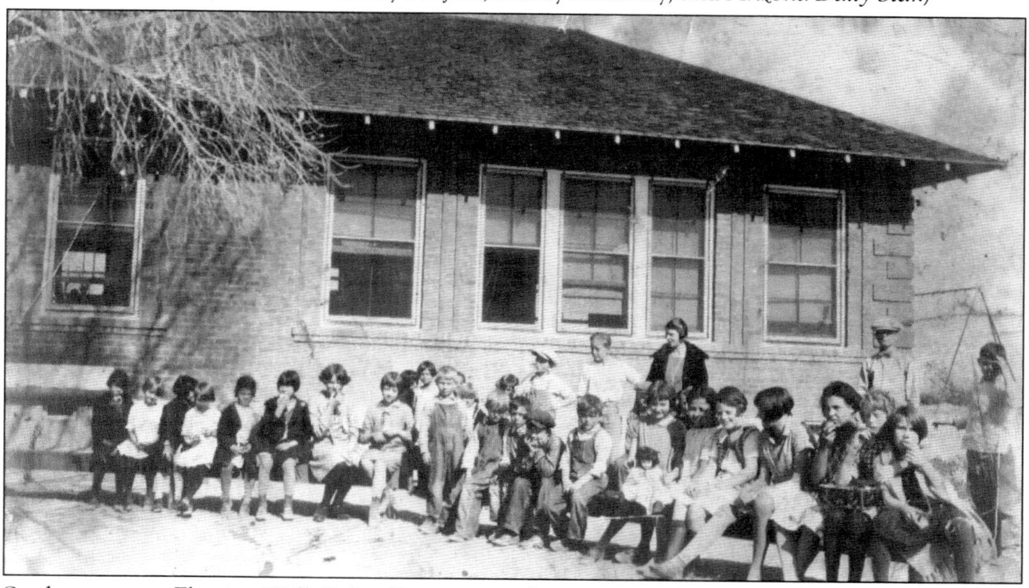

Students pose at Flowing Wells Schoolhouse around 1920. Around 1928, the district added a manual (vocational) training room, a manual training teacher, a kindergarten teacher, and a music teacher. When the Depression hit, Flowing Wells had to scale down, and manual training and music were dropped until better times in 1937, when two adjoining acres were purchased. Some teachers who are remembered include Mrs. Brown, Mrs. Bloom, and Bess Smith. (FWUSD and *Arizona Daily Star.*)

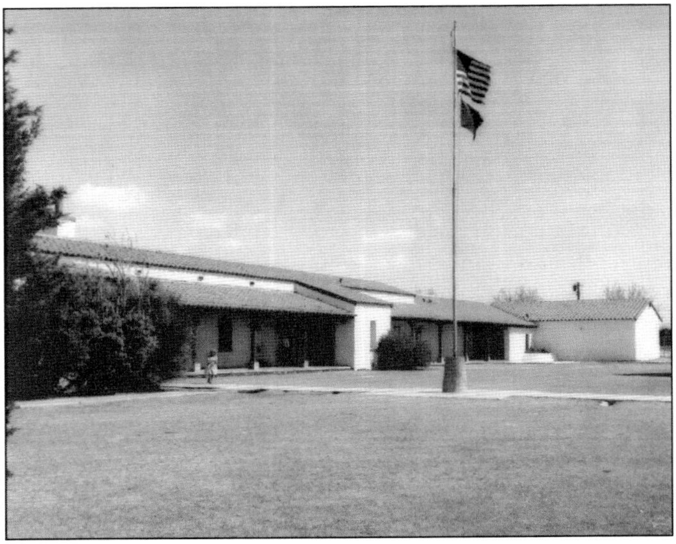

This building was constructed in 1940 with the help of the WPA for $15,000 at 1556 West Prince Road. Bonds provided $10,000, and $5,000 came from an insurance payment after the fire. The Spanish-style building with a red tile roof and corridor was designed by architect James McMillan. This building maintains its original character and is now the FWUSD administration building. School board meetings are held in the stage and auditorium area. (FWUSD and *Arizona Daily Star*.)

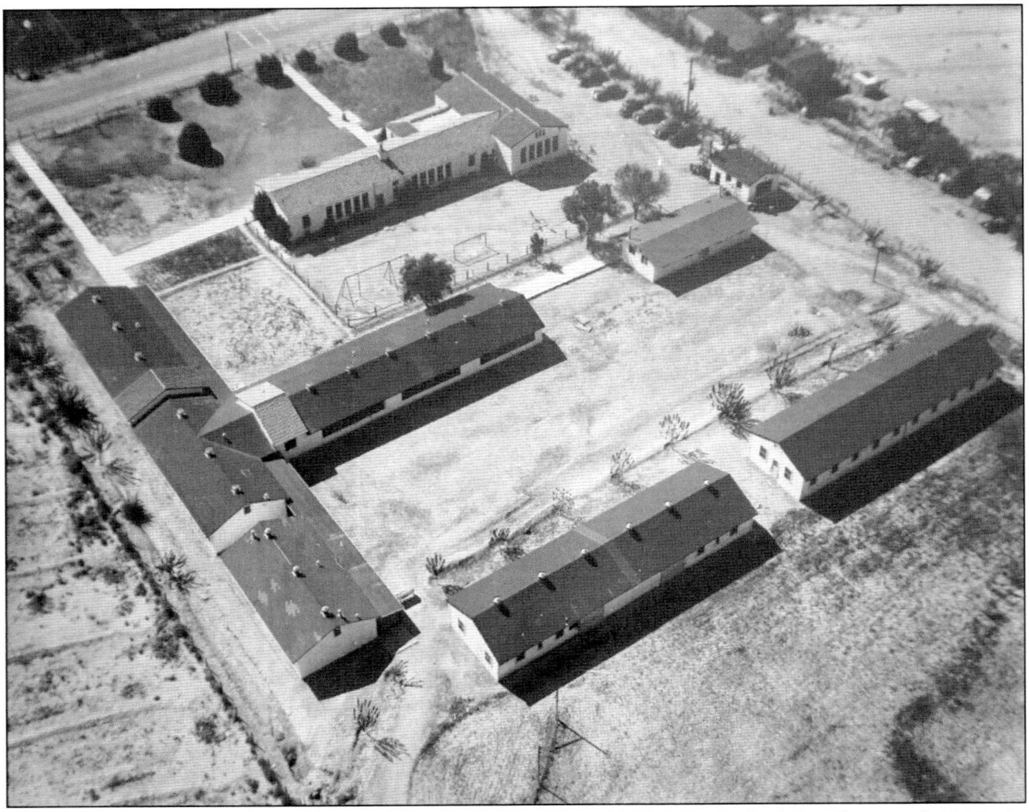

This is an aerial view looking southeast from the back of Flowing Wells School around 1949. At upper left, Prince Road was only two lanes and there was a large front lawn between it and the school building. The original building, finished in 1940, is at top center. Flowing Wells School District purchased surplus barracks (left and bottom) after World War II and converted them into classrooms to relieve overcrowding. Average daily attendance in 1942–1943 was 77, but by 1946–1947, it rose to 178. Janet Woods Elliott (FWHS 1959) remembers being frightened watching her cousins trying to swing the smaller kids high and "over" the swing set after school. (FWUSD.)

Iola Frans and Celia Gee were presented with certificates of life membership in the parent-teacher association at the Founder's Day banquet on February 6, 1956. Iola Frans (1910–1978) was the fourth teacher hired at Flowing Wells Elementary School, in 1940. She started out teaching second grade. During World War II, she became the acting principal. After receiving her master's degree in education from the University of Arizona, she was named the permanent principal in 1953. She was selected as Tucson Woman of the Year in Education and to Who's Who in American Women in 1967. As a founder of the Arizona School Administrators Association, Frans served as the group's first president in 1971. In her honor, Flowing Wells Elementary School was renamed Iola Frans School in 1956. She retired from the district in 1975 after 35 years of service. (FWUSD.)

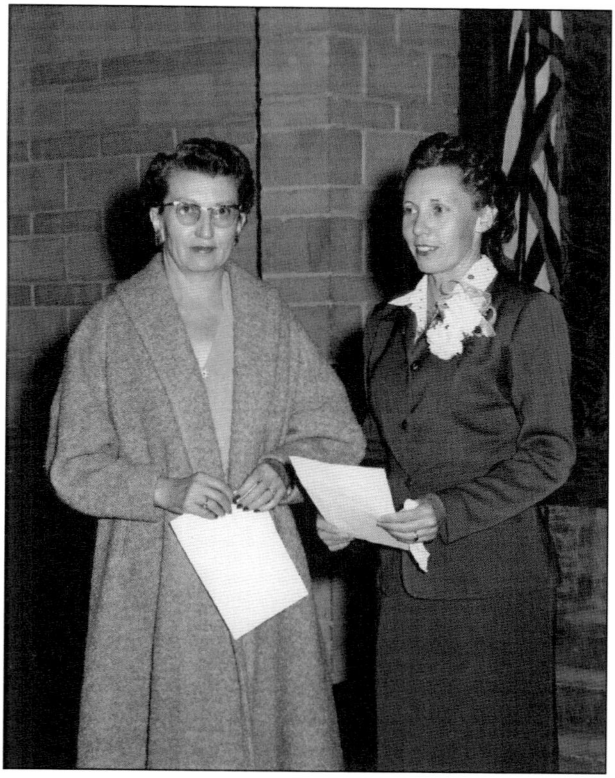

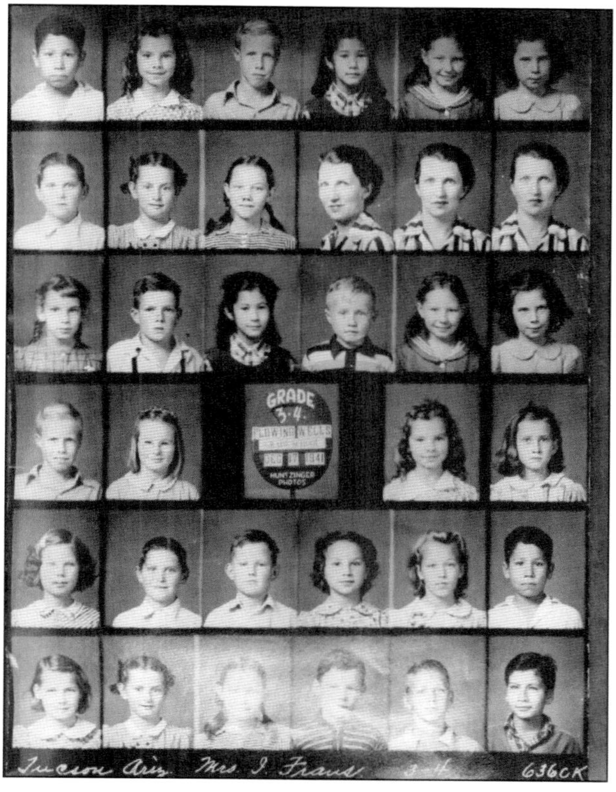

Students in Iola Frans's combined third- and fourth-grade classroom at Flowing Wells Elementary School are pictured in 1941. The five-classroom Flowing Wells Elementary School was built in 1940 with WPA labor. (FWUSD.)

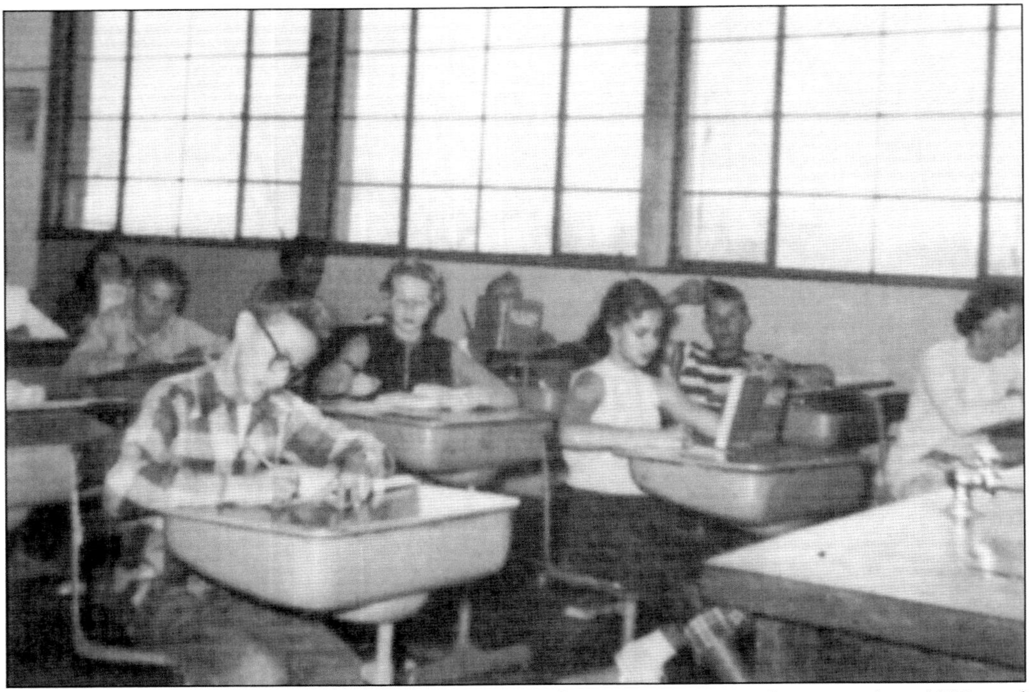

Shown is a 1954 class at Flowing Wells School. Iola Frans was named principal in 1953, and before her retirement in 1975, she held the position of assistant superintendent for educational services. (FWUSD.)

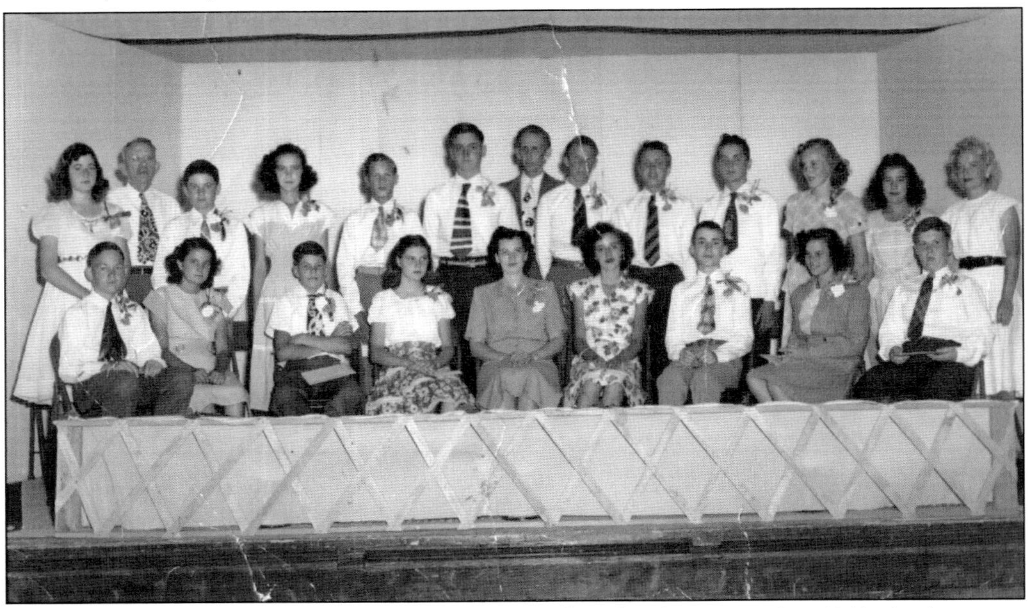

The eighth-grade graduation class from Flowing Wells School is pictured in 1948. At that time, Flowing Wells did not have a high school, so after graduation, students could choose between Amphitheater and Tucson High. Ninth grade was added to the Flowing Wells School in 1948, but the first full ninth-grade graduation ceremony was in 1952, as remembered by Marcia Claridge, who was in that class. The adults pictured here are Homer Davis (left), Herbert Barber (center, back), and Iola Frans (center, front). (FWUSD.)

Flowing Wells School 4-H boys and girls sang original songs during a 45-minute entertainment program on March 5, 1940, over KVOA radio in Tucson. The students are believed to be, from left to right, Manuel Lopez, Clair Dobias, Loretta Gray, and Josephine Lopez. The man holding the microphone is unidentified. (FWUSD.)

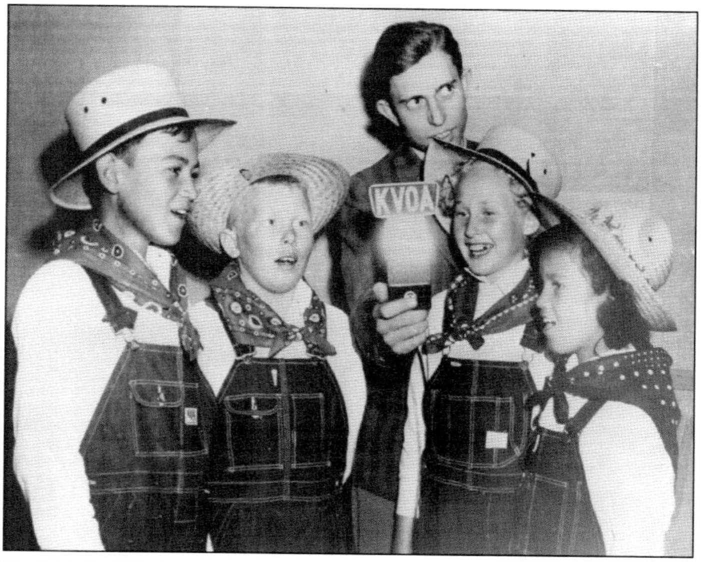

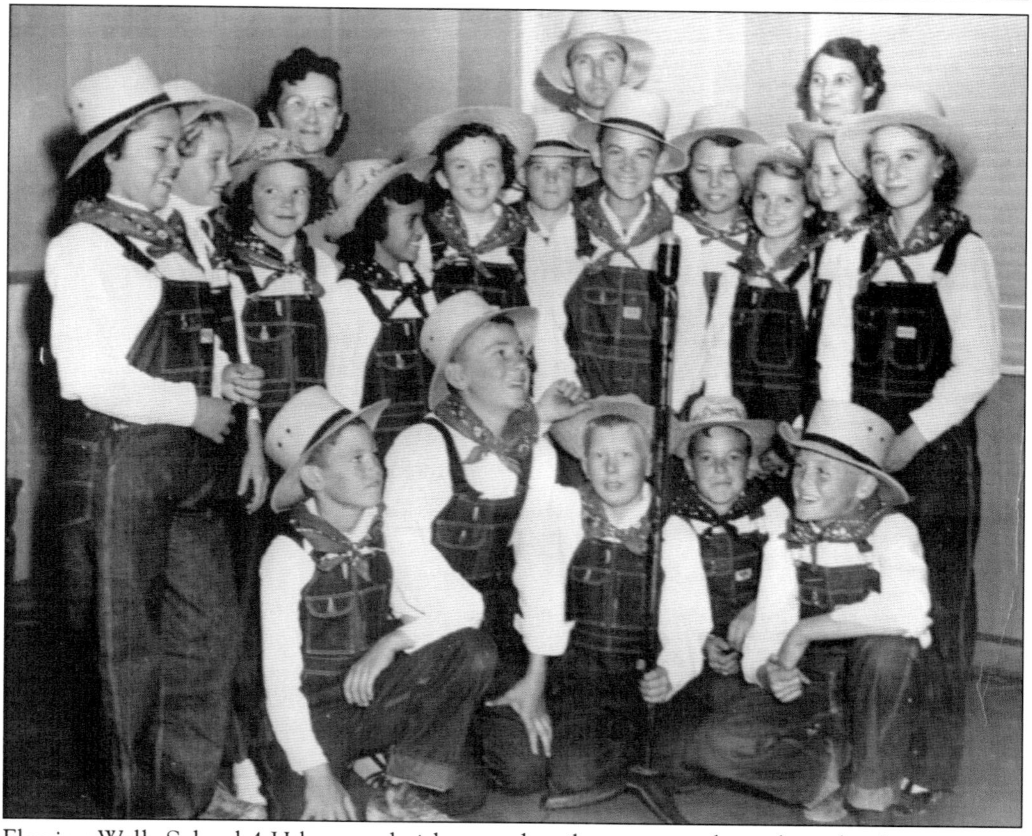

Flowing Wells School 4-H boys and girls won the silver cup on the rodeo school program on February 23, 1940. From left to right are (kneeling) Bobby Honea, Sam Marlar, Clair Dobias, Sidney Pittman, and Edward Phillippe; (standing) Elsie Ortega, Loretta Gray, Nettie May Pittman, Josephine Lopez, Mary McAnally, Tillman Pittman, Manuel Lopez, Eva Ortega, Joe Ann Kine, Jane Criger, and Ellen McSpadden. The leaders, in back, are Jane Copeland, Ed Van Doren, and Ruth Downs. (FWUSD.)

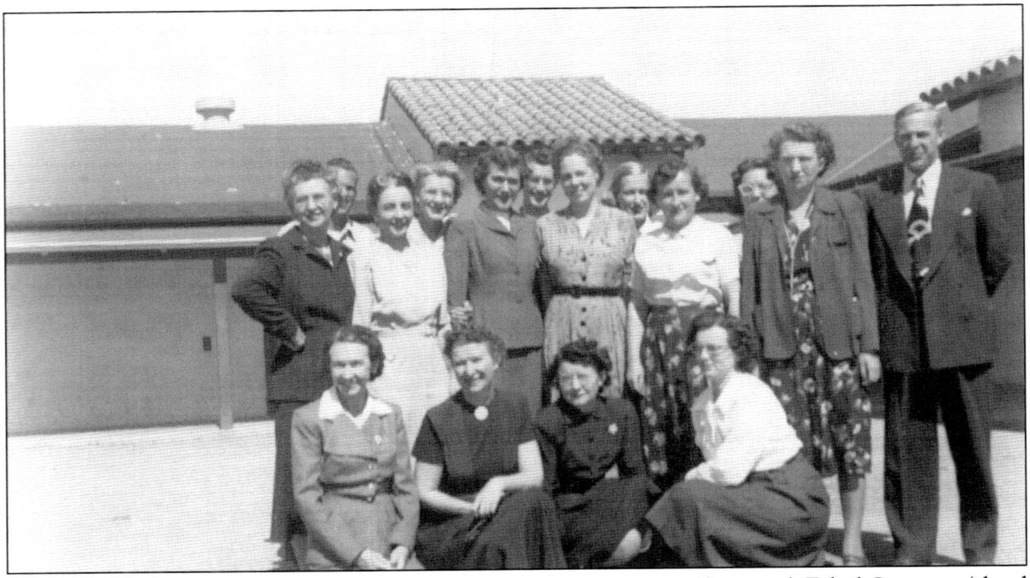

Teachers at Flowing Wells in 1952–1953 are, from left to right, (first row) Ethel Seaman (third grade), Irene Tompkins (first grade), Olive Harris (second grade), Mrs. G. Kindred (ninth grade), Ann Brown (third grade), Ethyl Futrell (fifth grade), Ruth Priebe (fifth grade), Gladys Taylor (fourth grade), Wanda Adams (first grade), Iola Frans (second grade), and Ralph Roda (principal); (second row) A.C. Williams (social studies and coach), Helen Hendricks (fourth grade), Duane Miller (music), Mrs. L. Allen (fourth grade and art), and Alice Tremblay (seventh grade and English). Beloved teacher Irene Tompkins's 100th birthday celebration in 2004 was attended by her students. She came to Flowing Wells with 13 years of teaching experience in 1942 and retired in 1969. Randy Holladay (FWHS 1960) remembered that she drew huge murals in pastels on her blackboard in his first-grade class in 1948. (FWUSD.)

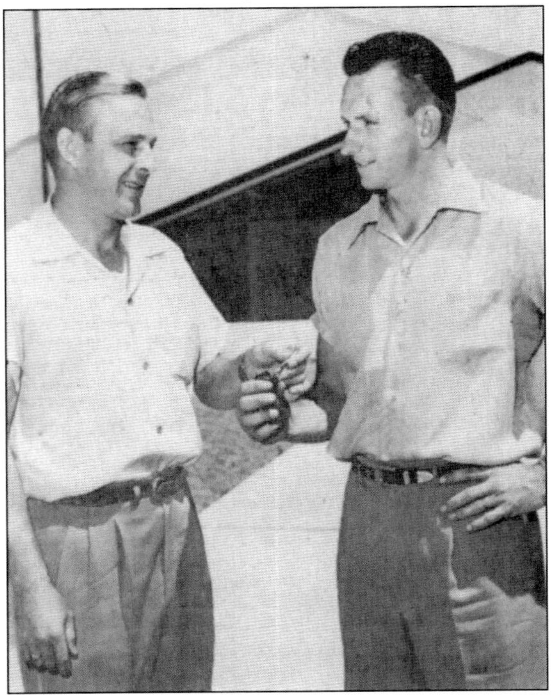

Ralph Roda came to Tucson in 1949 and became the principal and superintendent of Flowing Wells in 1950. In his first year at Flowing Wells, there were 350 students and 10 teachers. The enrollment in 1951 increased to 515, with a teacher staff of 20. Under Roda's administration, nine new classrooms, new administrative offices, and a multipurpose building were added in 1953. Roda is shown passing the keys to Flowing Wells to the new superintendent, George N. Smith, on July 1, 1953. (FWUSD.)

FACULTY

Ralph W. Roda, *Principal*

Mrs. Kindred	*Junior High*
Miss Tremblay	*Junior High*
Mr. Barber	*Junior High*
Mr. Aguilu	*Junior High*
Mr. Williams	*Physical Education*
Mr. Miller	*Music*

Graduation Exercises

FLOWING WELLS
JUNIOR HIGH SCHOOL
DISTRICT NO. 8

WEDNESDAY EVENING, MAY 28, 1952
EIGHT O'CLOCK

Pictured is the cover of the commencement program for the first ninth-grade graduation exercise at Flowing Wells Junior High on May 28, 1952. At the time, kindergarten through ninth-grade students attended one location on Prince Road. This was one year before the completion of the new multipurpose building, which is now the athletics building at FWHS. (FWUSD.)

GRADUATES

Sally I. Baillie	Eugene M. Radza
Donna L. Baughman	John R. Shea
Susan D. Berry	Victor R. Shelton
James D. Cloakey	Edward L. Sherfey
David D. Dobias	Billie M. Simington
Joan C. Farrelly	DeLois A. Sipe
Marcia Gardner	Russell G. Steinebach
Norris E. Gompf	Patricia R. Stoddard
Rathie A. Harbort	Audrey J. Stoll
Patricia R. La Vallee	John S. Swaim
Michael F. Meyer	Joan E. Tenney
Grace M. Nunez	Thomas E. Tyler
James A. Peery	James M. Vanderkolk
Aline N. Porter	Joyce Y. Ward

USHERS

Lannis Smith	William Stallings
Clara Rice	James Whitlock

PROGRAM

Processional	Band
Invocation	Rev. E. A. Hoefer
Salutatorian	John Swaim
"Get On Board, Little Children" Negro Spiritual	Class
Class History	Marcia Gardner
Class Prophesy	Patsy Stoddard
"Whispering Hope" - Alice Hawthorne	Class
Valedictory	Sally Baillie
Class Will	Eugene Radza
"Now The Day Is Over" - Barnaby	Class
Address to the Graduates	Colonel William C. Sipes, Jr.
American Legion Constitution Award	Holden Olsen
American Legion Citizenship Award	Holden Olsen
Perfect Attendance Awards	Ralph W. Roda
Athletic Awards	A. C. Williams
Band Awards	Duane Miller
Presentation of Diplomas	Irven E. Gee
Patrol Awards	Sergeant Arthur Grande
Benediction	Rev. E. A. Hoefer
Recessional	Band

Pictured is the inside of the program for the ninth-grade graduation on May 28, 1952. Ralph Roda was the principal. (FWUSD.)

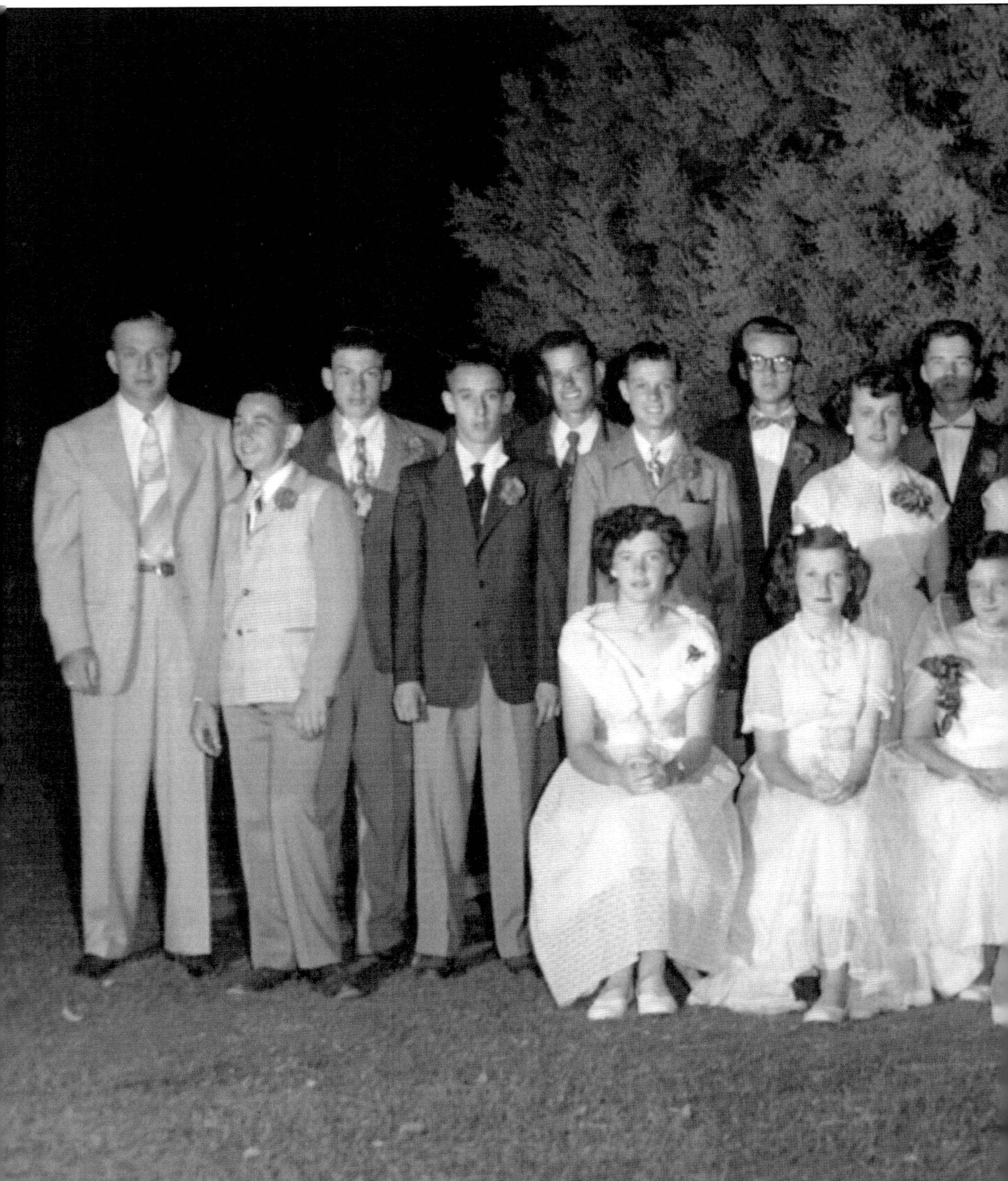

This is the Flowing Wells Junior High ninth-grade graduating class of 1952. Since Flowing Wells did not have a high school, students had to choose to attend Amphitheater or Tucson High. In this class, Patsy Stoddard, Marie Simington, and David Dobias are the only students who attended from first through ninth grades. From left to right are (first row) Joan Farrelly, Susan Berry, Sally Ballie, Donna Baughman, Patsy Stoddard, Patty LaVallee, Audrey Stoll, and Joyce Ward;

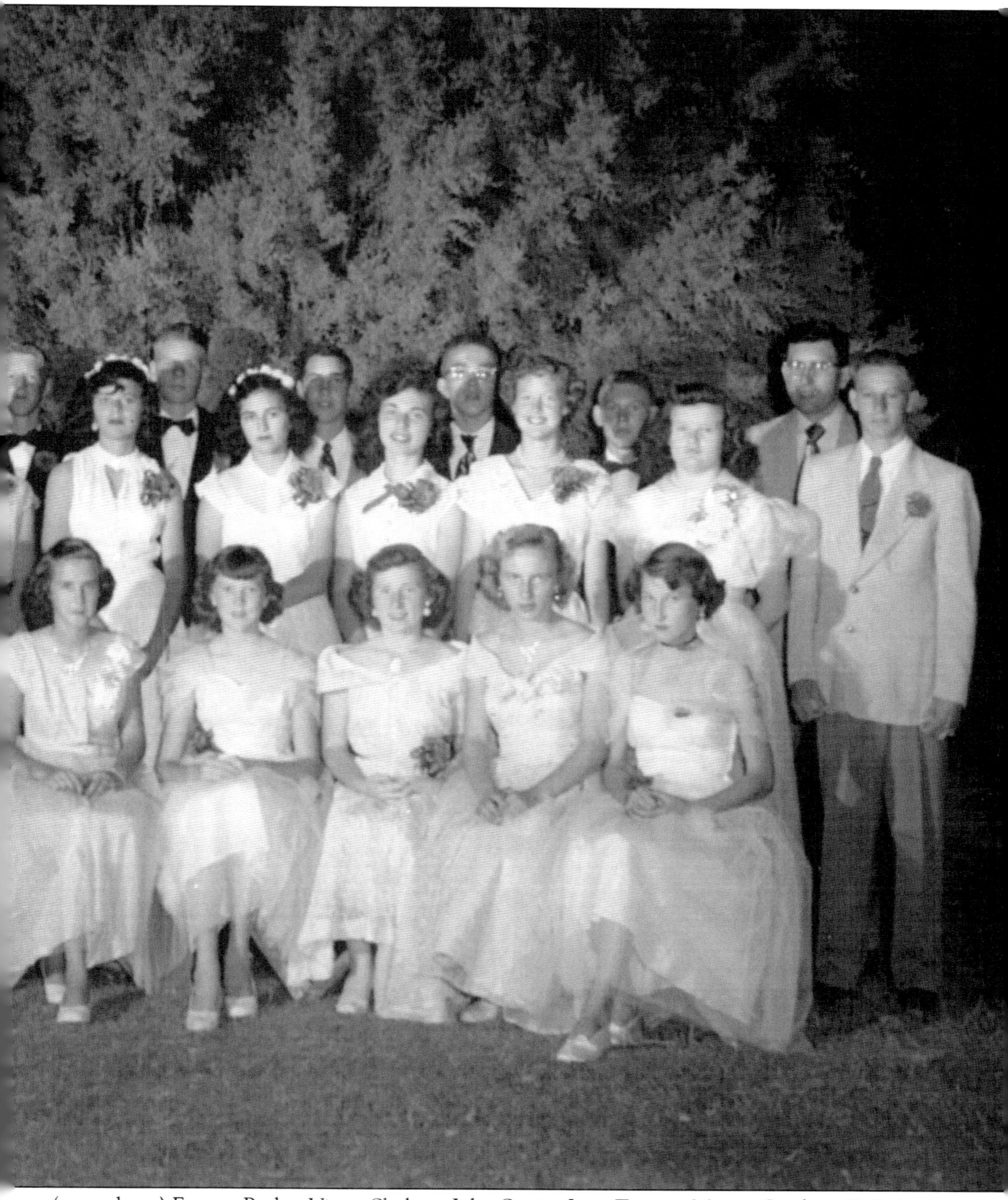

(second row) Eugene Radza, Victor Shelton, John Swain, Joan Tenney, Marcia Gardner, DeLois Sipe, Grace Nunez, Rathie Ann Harbort, Marie Simington, Aline Porter, and Eddie Sherfey; (third row) Principal Ralph Roda, Jim Peery, Eugene Tyler, John "Spec" Shea, Jim Cloakey, Norris Gompf, David Dobias, Mickey Meyer, Jim Vanderkolk, Russell "Buster" Steinbach, and Julian Aguilu (science and math). (Marcia Gardner Claridge and FWUSD.)

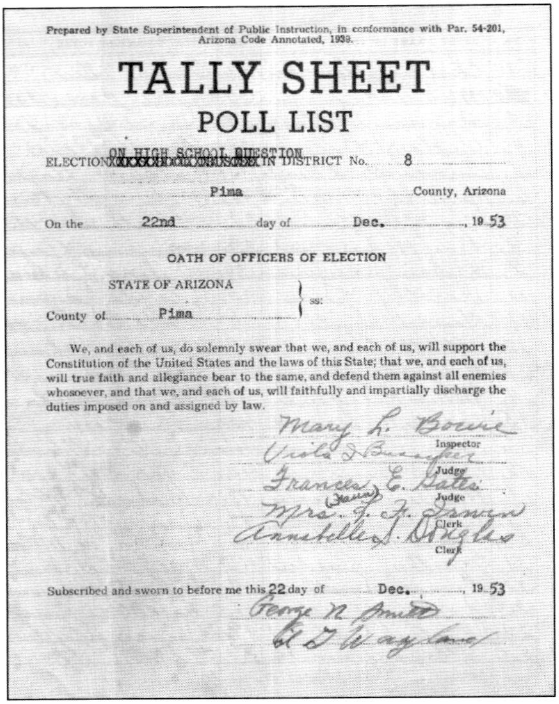

Prepared by State Superintendent of Public Instruction, in conformance with Par. 54-201, Arizona Code Annotated, 1939.

TALLY SHEET

POLL LIST

ELECTION ON HIGH SCHOOL QUESTION IN DISTRICT No. 8

Pima County, Arizona

On the 22nd day of Dec., 19 53

OATH OF OFFICERS OF ELECTION

STATE OF ARIZONA }
County of Pima } ss:

We, and each of us, do solemnly swear that we, and each of us, will support the Constitution of the United States and the laws of this State; that we, and each of us, will true faith and allegiance bear to the same, and defend them against all enemies whosoever, and that we, and each of us, will faithfully and impartially discharge the duties imposed on and assigned by law.

Mary L. Bowie, Inspector
Viola D Buspecker, Judge
Frances E. Gates, Judge
Mrs. F. F. (Fauna) Irmen, Clerk
Annabelle S. Douglas, Clerk

Subscribed and sworn to before me this 22 day of Dec., 19 53

George N. Smith
E. D. Wayland

On December 22, 1953, the Flowing Wells community voted to create Flowing Wells High School District No. 8. Flowing Wells is Tucson's fourth high school. Previously there was only Flowing Wells Elementary School District No. 8. This is the front page of the vote tally sheet approving FWHS. (Geraldine Gee Larrington.)

The tally sheet shows 63 for and three against creating the Flowing Wells High School District No. 8. Following this successful vote, the Flowing Wells community approved the sale of $340,000 in school improvement bonds on May 8, 1954—$260,000 was to build FWHS, and $80,000 was for the elementary school. In 1977, the Flowing Wells Elementary and High School Districts became one unified district, simplifying finances and paperwork. (Geraldine Gee Larrington.)

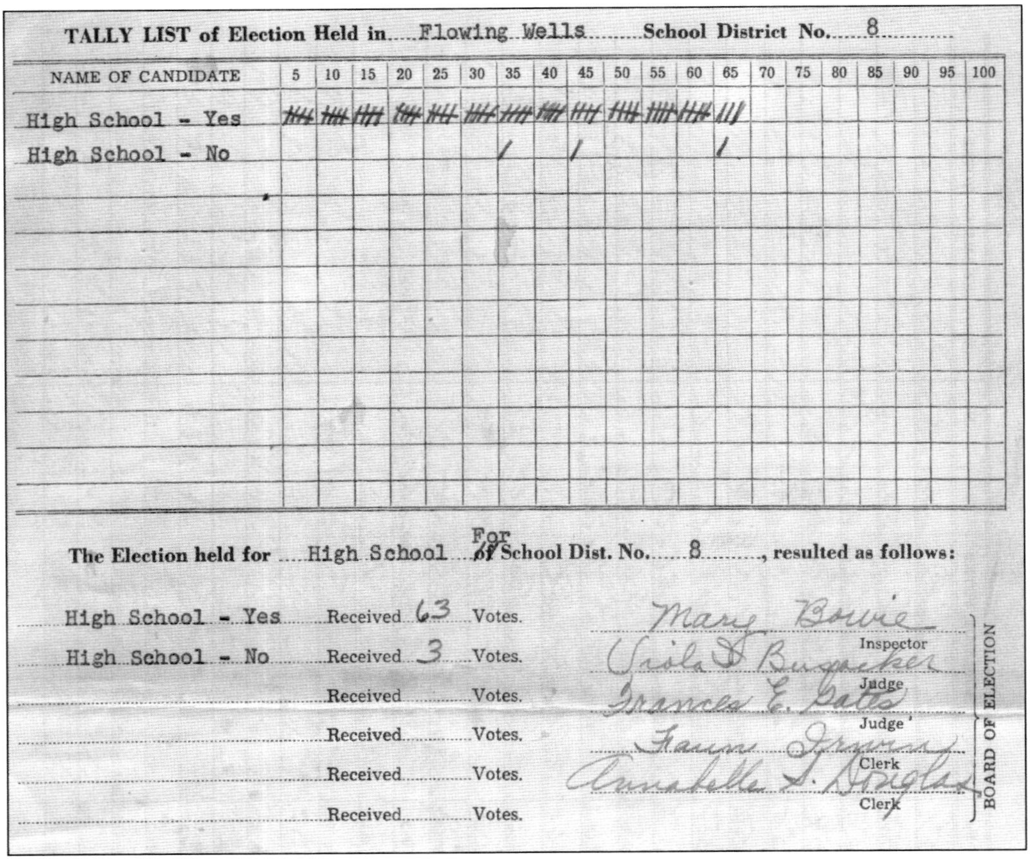

TALLY LIST of Election Held in Flowing Wells School District No. 8

NAME OF CANDIDATE	5	10	15	20	25	30	35	40	45	50	55	60	65	70	75	80	85	90	95	100
High School - Yes	𝍸	𝍸	𝍸	𝍸	𝍸	𝍸	𝍸	𝍸	𝍸	𝍸	𝍸	𝍸	///							
High School - No							/		/				/							

The Election held for High School For School Dist. No. 8, resulted as follows:

High School - Yes Received 63 Votes.
High School - No Received 3 Votes.
Received Votes.
Received Votes.
Received Votes.
Received Votes.

BOARD OF ELECTION:
Mary Bowie, Inspector
Viola D Buspecker, Judge
Frances E. Gates, Judge
Fauna Irmen, Clerk
Annabelle S. Douglas, Clerk

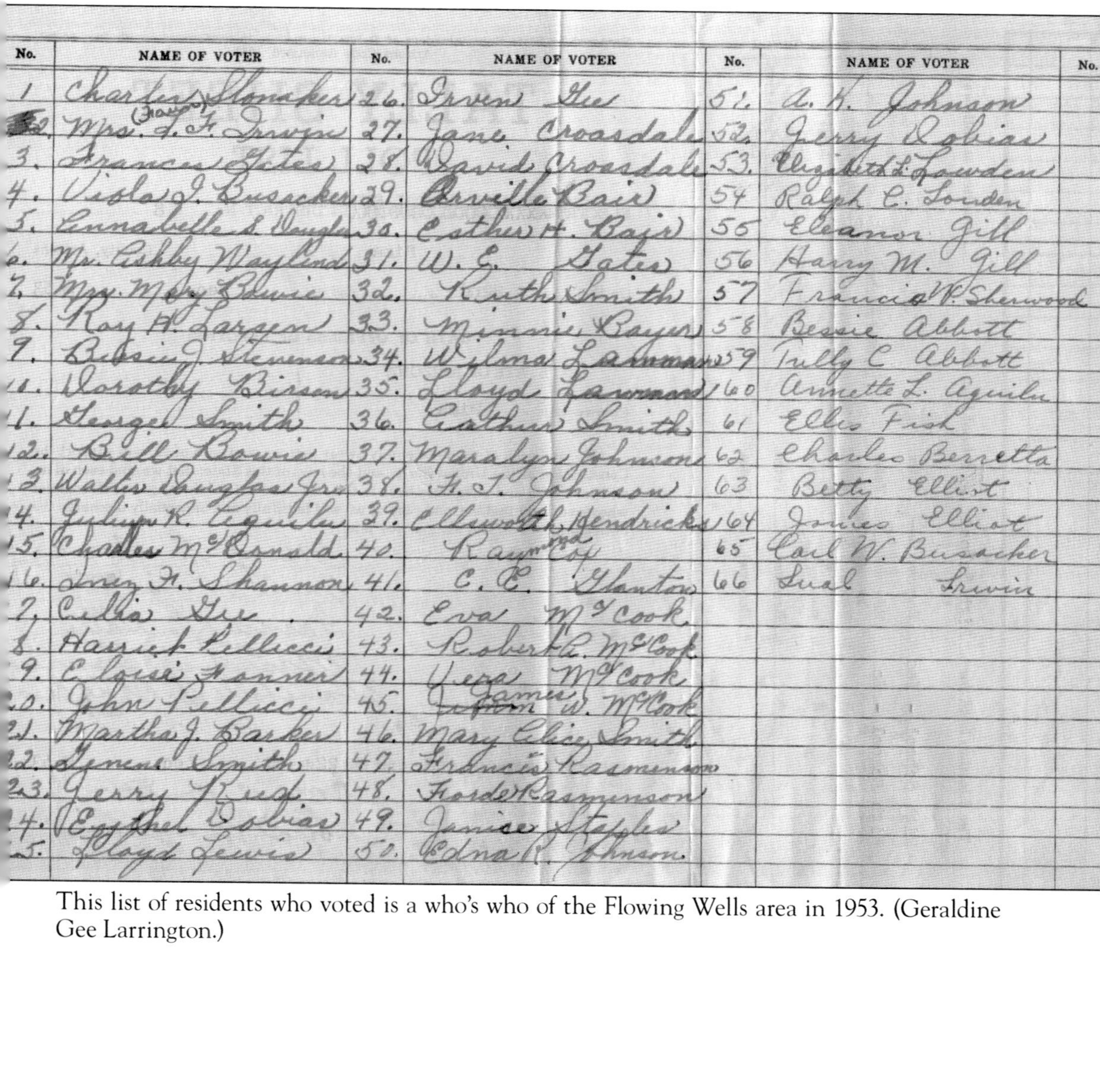

No.	NAME OF VOTER	No.	NAME OF VOTER	No.	NAME OF VOTER	No.
1	Charles Slonaker	26.	Irven Gee	51.	A. K. Johnson	
2	Mrs. (Frances) L. F. Irvin	27.	Jane Croasdale	52.	Jerry Dobias	
3.	Frances Gates	28.	David Croasdale	53.	Elizabeth L. Louden	
4.	Viola J. Busacker	29.	Orville Bair	54	Ralph C. Louden	
5.	Annabelle S. Douglas	30.	Esther H. Bair	55	Eleanor Gill	
6.	Mr. Ashley Wayland	31.	W. E. Gates	56	Harry M. Gill	
7.	Mrs. Mary Bowie	32.	Ruth Smith	57	Francis W. Sherwood	
8.	Roy H. Larsen	33.	Minnie Bayer	58	Bessie Abbott	
9.	Bessie J. Stevenson	34.	Wilma Lawman	59	Tully C Abbott	
10.	Dorothy Birsen	35.	Lloyd Lawman	60	Annette L. Aguilar	
11.	George Smith	36.	Cathrine Smith	61	Ellis Fish	
12.	Bill Bowie	37.	Maralyn Johnson	62	Charles Berretta	
13.	Walter Douglas Jr.	38.	H. T. Johnson	63	Betty Elliot	
14.	Julian R. Aguilar	39.	Ellsworth Hendricks	64	James Elliot	
15.	Charles McDonald	40.	Raymond Coy	65	Carl W. Busacker	
16.	Inez F. Shannon	41.	C. E. Glanton	66	Leal Irvin	
17.	Cellia Gee	42.	Eva McCook			
18.	Harriet Pellicci	43.	Robert A. McCook			
19.	Eloise Fanner	44.	Vera McCook			
20.	John Pellicci	45.	James W. McCook			
21.	Martha J. Barker	46.	Mary Alice Smith			
22.	Genene Smith	47.	Francis Rasmenson			
23.	Jerry Reed	48.	Flord Rasmenson			
24.	Esther Dobias	49.	Janice Staples			
25.	Lloyd Lewis	50.	Edna R. Johnson			

This list of residents who voted is a who's who of the Flowing Wells area in 1953. (Geraldine Gee Larrington.)

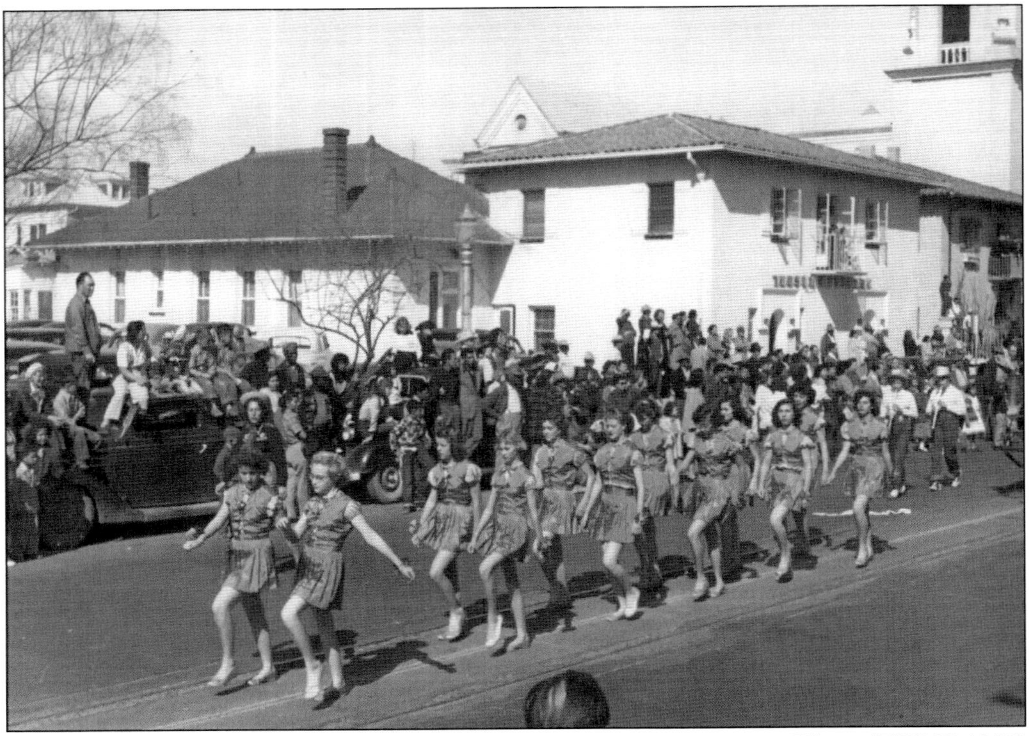

The FWHS band marches in the rodeo parade in February 1954. Jim Elliott (FWHS 1958) remembers that the lead group was the majorettes, then the band, then the marching girls. It is believed that the outfits the girls are wearing were made by a parent. Mary Jane Brown (FWHS 1958) said that the shoes the girls are wearing were ballet slippers. She said, "I wonder how long my feet hurt after that." Jon Elliott (FWHS 1960) recalled that the band and the girls put on a little show along the route to the tune of "Don't Fence Me In." He said, "they changed from march formation to a big square box to emulate a corral fence." He thinks the band "fenced in" the marching girls. Mary Jane Brown is in the first row facing forward on the left; Jean Woods O'Hanlon is at right. Directly behind Jean is Elizabeth Chase. Next to Elizabeth is believed to be Sherry Gates Brooks. (FWUSD.)

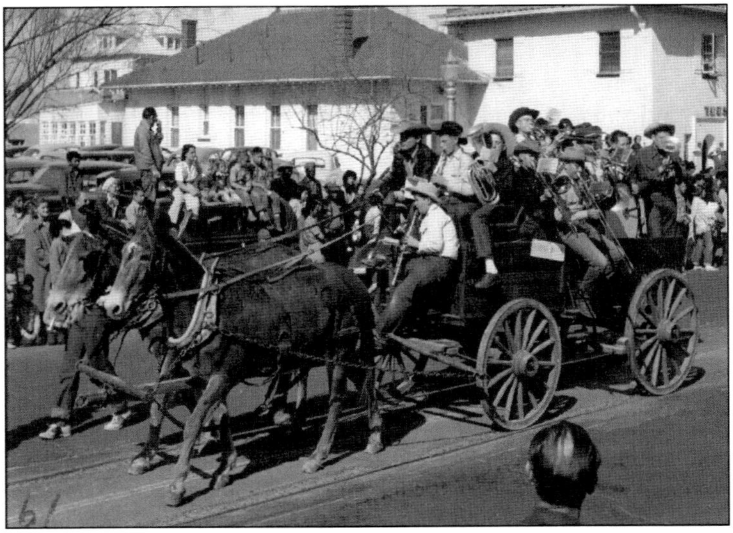

The FWHS band rides and performs on a horse-drawn wagon in the rodeo parade in 1961. The first band was formed at Flowing Wells School in 1951. (FWUSD.)

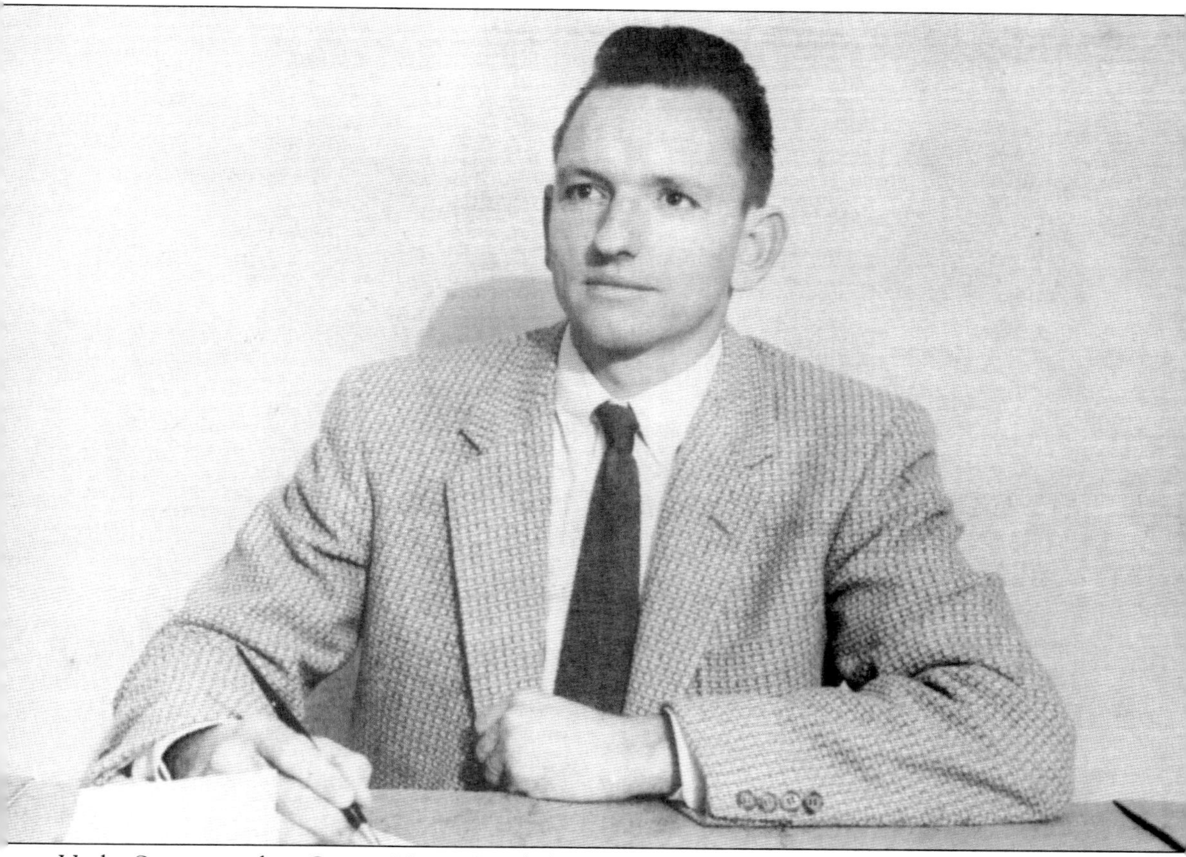

Under Superintendent George "Spec" Smith (1953–1967), Flowing Wells grew from one school, 20 teachers, and 500 students to five schools, 148 teachers, and 3,400 students. At 26, Smith came to Flowing Wells as the youngest superintendent of a major school district in Arizona. He was born in 1927 in Safford, Arizona. He was an Army paratrooper, and married Mary Alice and had four children: Diana Thomas (FWHS 1966), Jennifer Johnson (FWHS 1967), David, and Scott. George played basketball for the University of Utah, graduated, and became head teacher and principal in Arlington, Arizona. After Flowing Wells, he became superintendent of Mesa Unified for 18 years, a member of the Arizona State Board of Education for six years, and founding president of the Arizona School Administrators Association, and was appointed twice by President Reagan to the President's Intergovernmental Advisory Council on Education. He completed a master's degree from Arizona State University and a doctorate of education from Arizona State University and Stanford. As executive vice president at Northern Arizona University, he established satellite campuses around the state. Subsequent superintendents include William K. Poston (1974–1983), Dr. J. Robert Hendricks (1983–1996), Dr. Phillip M. Corkill (1996–1998), Dr. John J. Pedicone (1998–2004), Dr. Nicholas I. Clement (2004–2013), and Dr. David R. Baker (2013–present). (FWUSD.)

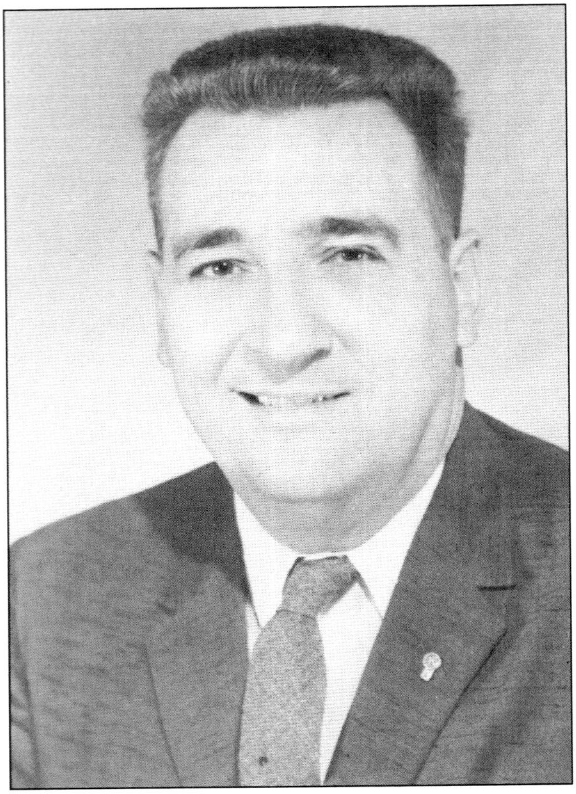

William "Bill" Lovett served as assistant superintendent (1956–1966) and director of secondary education. Lovett was the first principal of FWHS from 1956 to 1966. He was instrumental in bringing accredited academics to Flowing Wells. Lovett was responsible for compiling the statistics and records that won Flowing Wells the 26th annual Bellamy Award in 1967. The Bellamy Award is presented in one state each year to honor Francis Bellamy (the writer of the Pledge of Allegiance) and to foster patriotism among young people. FWHS was presented with the award on October 13, 1967. (FWUSD.)

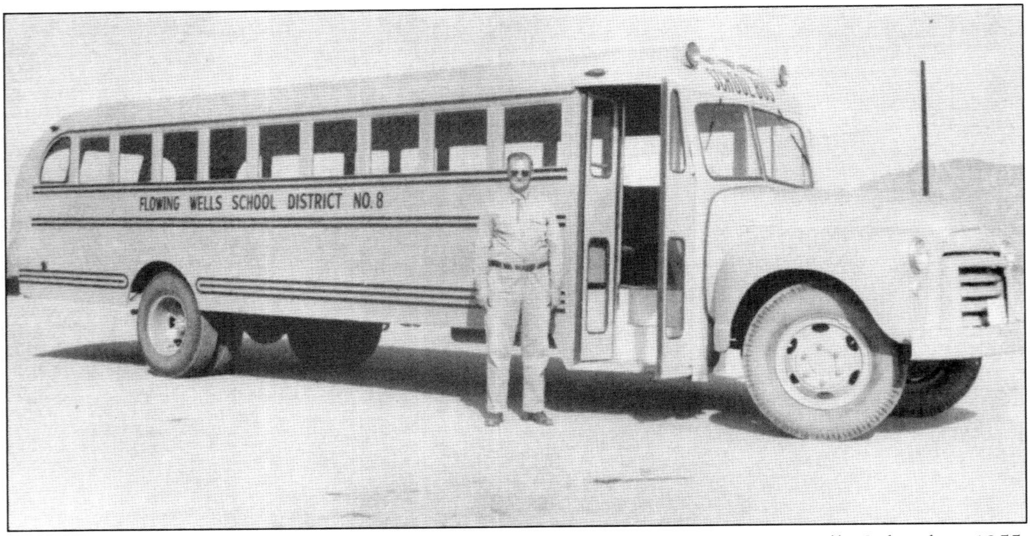

Vernon Soule was a maintenance man and school bus driver at Flowing Wells School in 1955. Soule moved to Tucson in 1952 and purchased land on Prince Road where the Green Meadows Trailer Park now sits. When the school got a new bus, Soule went to Michigan to drive it back to Flowing Wells. His granddaughter Lynette Smith Christian (FWHS 1965 and FWHS Athletics Hall of Fame's first class, 2003) said, "I always made sure I was on my grandfather's bus to games, field trips, and events." Soule delivered kids to their destinations from 1954 to 1970. Ashby G. Wayland also drove the bus in 1954. (FWUSD.)

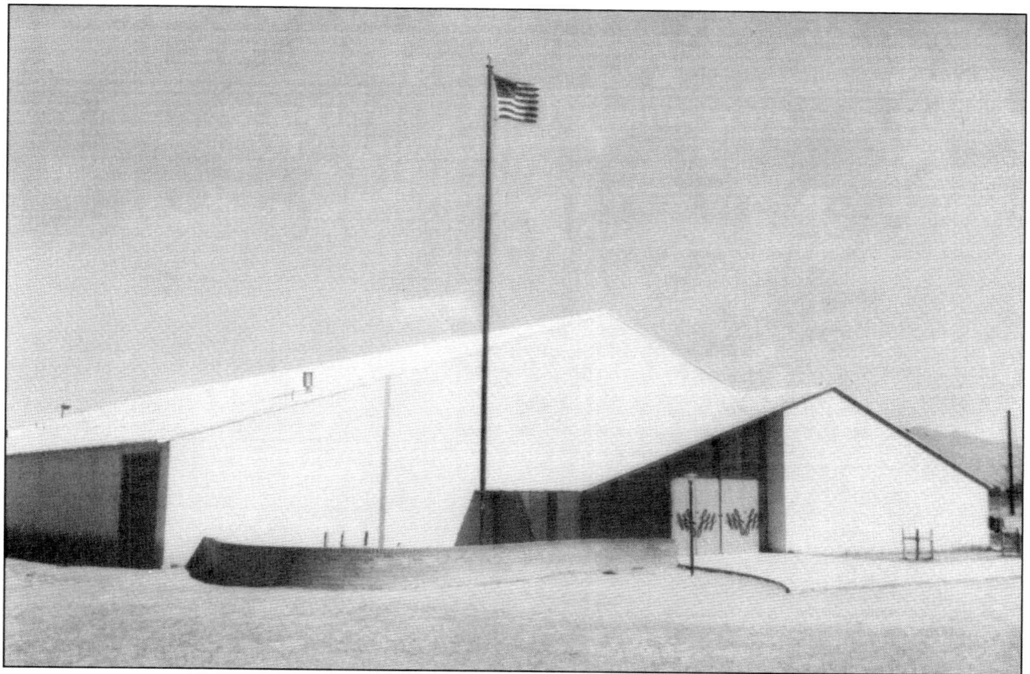

On May 27, 1953, Flowing Wells Junior High held commencement exercises for the ninth grade in the new multipurpose building (currently the athletics building). It cost $113,650 to complete. The building housed basketball, showers, a stage, and home economics classrooms. At the time, the entrance was to the south. (FWUSD.)

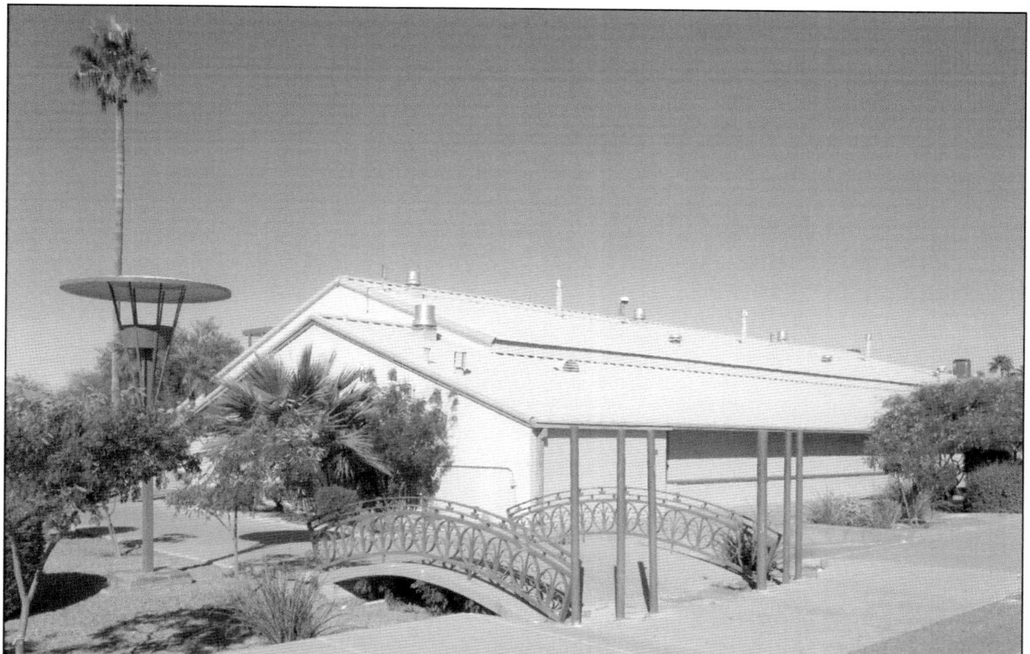

On the southeast corner, there is still today a concrete bridge and decorative light that provide passage over a former irrigation canal. Classmates of the time say this was a gathering spot, a popular picture spot, and a good place "to get your first kiss." (FWUSD.)

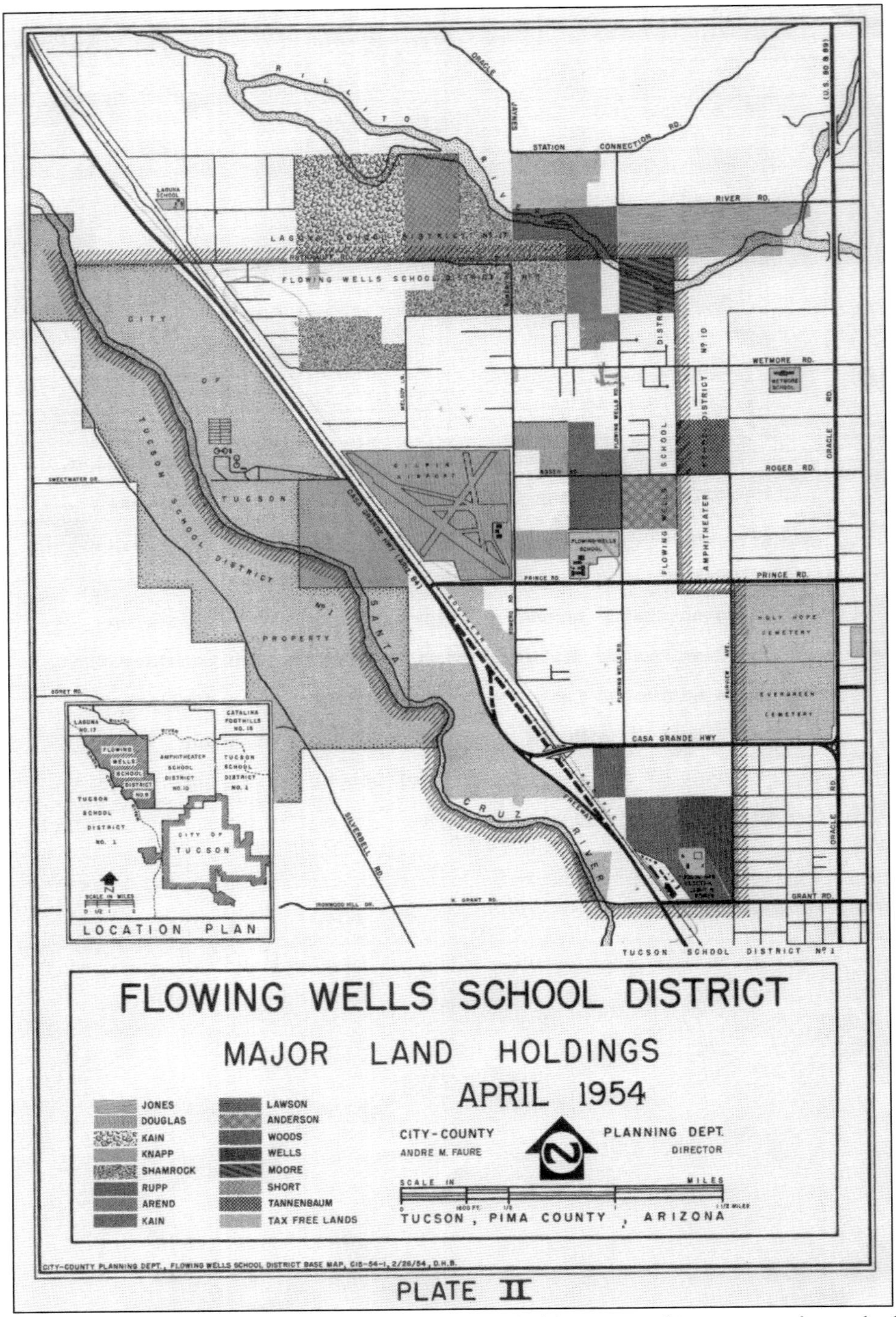

This Flowing Wells School District map of area land holdings in 1954 was part of a study done by the Tucson and Pima County Planning Departments to project future growth and student enrollment. Amphitheater School District was cut from the Rillito School District in 1893. Prince Road was named after Levi Marston Prince, a miner and homesteader who was a founding board member of Amphitheater School District. Miracle Mile was named by Stanley Williamson after the Miracle Mile in Los Angeles. (FWUSD.)

On March 20, 1958, Romero Road School was renamed Homer Davis Elementary in honor of the man who dedicated 52 years to teaching. In 1903, Davis moved to Arizona, where he taught for 20 years. He moved to Chicago in 1923 and retired in 1945. Faced with a teacher shortage, Davis came out of retirement in 1946 to assume the role of principal of Flowing Wells School until his second retirement in 1950 at the age of 70. He was the executive secretary of the Arizona School Board Association until 1958. (FWUSD.)

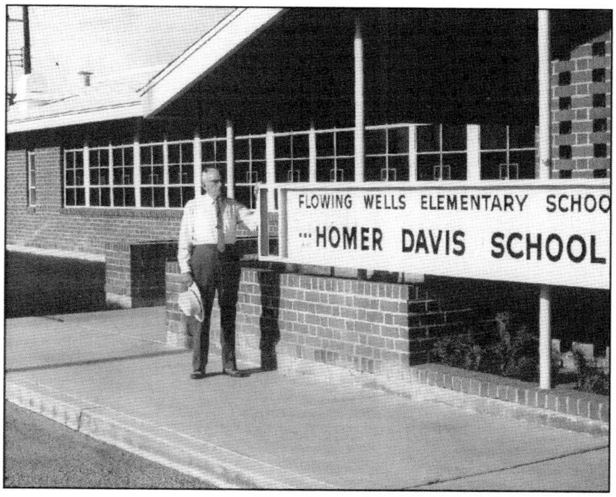

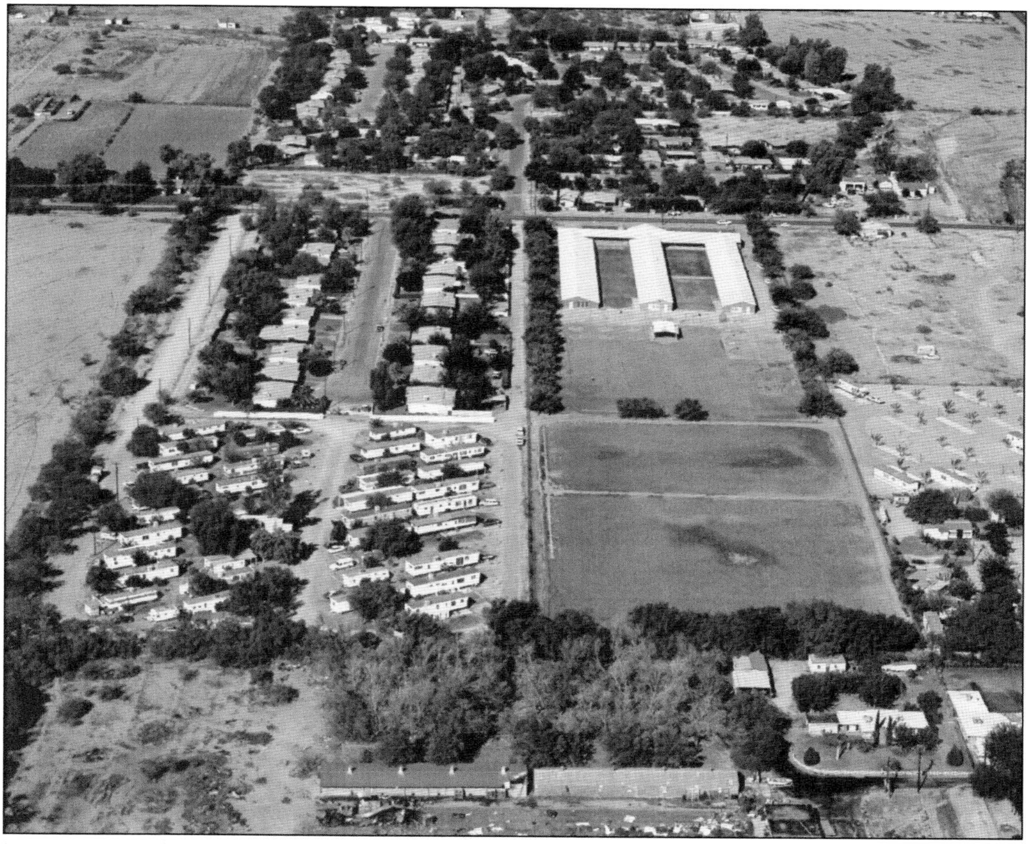

Homer Davis Elementary opened as Romero Road School in January 1956 with six classrooms. An additional six classrooms, an administration office, and a multipurpose room were completed in 1957. The school, pictured around 1965, opened with kindergarten through third grade. Root Lane is directly to the left of the school. The 22 homes on Root Lane produced about 32 top athletes between 1967 and 1985 with over 500 wins, including four award-winning starting quarterbacks: Richard Galvez Jr. (FWHS 1968), David Galvez (FWHS 1974), Steve Torres (FWHS 1978), and Tim Torres (FWHS 1985). Vicky Anzaldua (FWHS 1973) played semipro women's softball. Physical education teacher and coach Rhett Wilbur (1962–1996) said, "Every blooming flower has a root. The Flowing Wells Athletics Program had Root Lane." (FWUSD.)

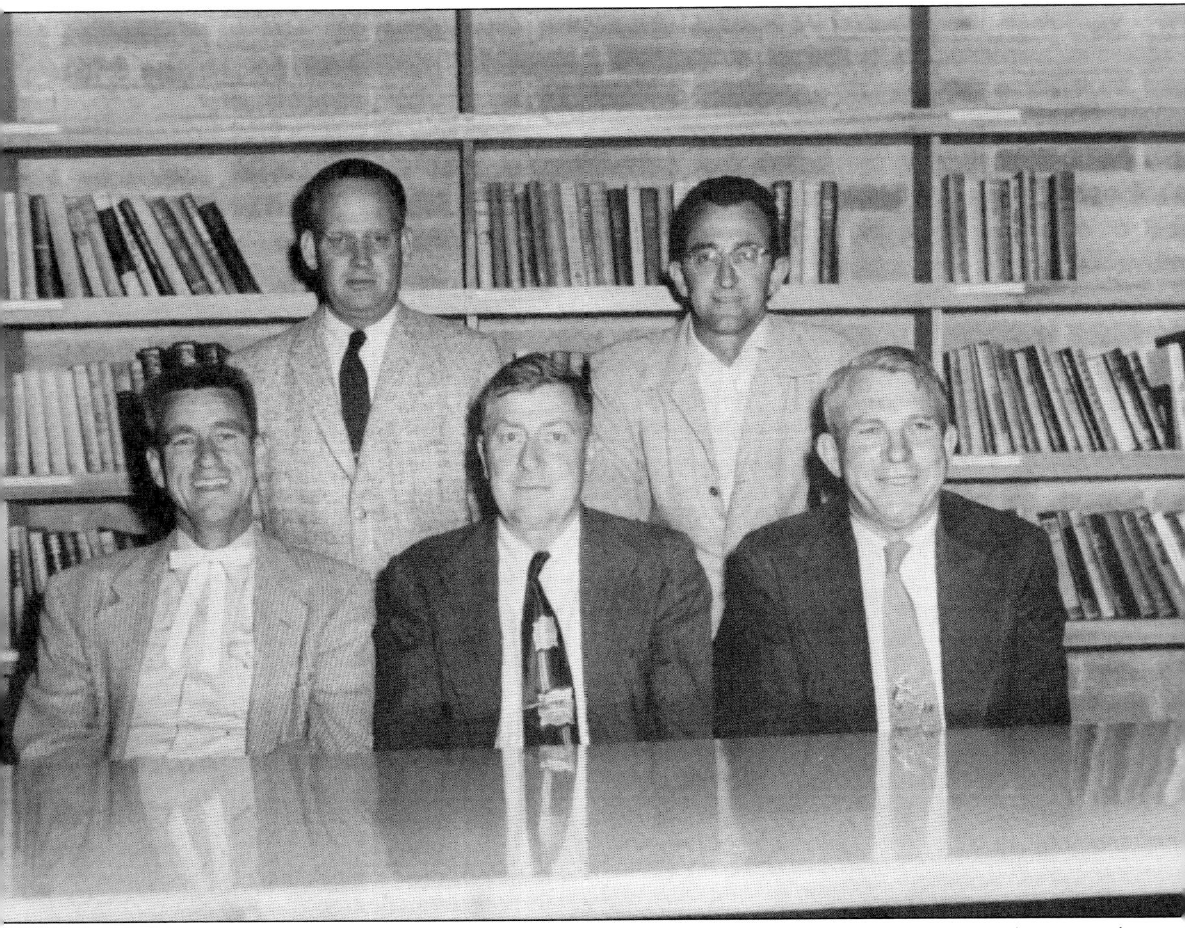

Board members are, from left to right, (first row) Carlos J. LaValle, Pres. Walter Douglas Jr., and Irven E. Gee; (second row) clerk Leonard Woods and Harry M. Gill. Woods and Gill joined in 1957 to form the first five-member Flowing Wells School Board. The first board under the Flowing Wells name in 1928 is believed to have included Henry Peterson, clerk; Mrs. E.C. Edwards, trustee; and Mrs. R.M. McKinney, trustee. There is also a reference to a Mr. C.F. Lovitt at that time, but he has not been confirmed as a trustee in school district records. (FWUSD.)

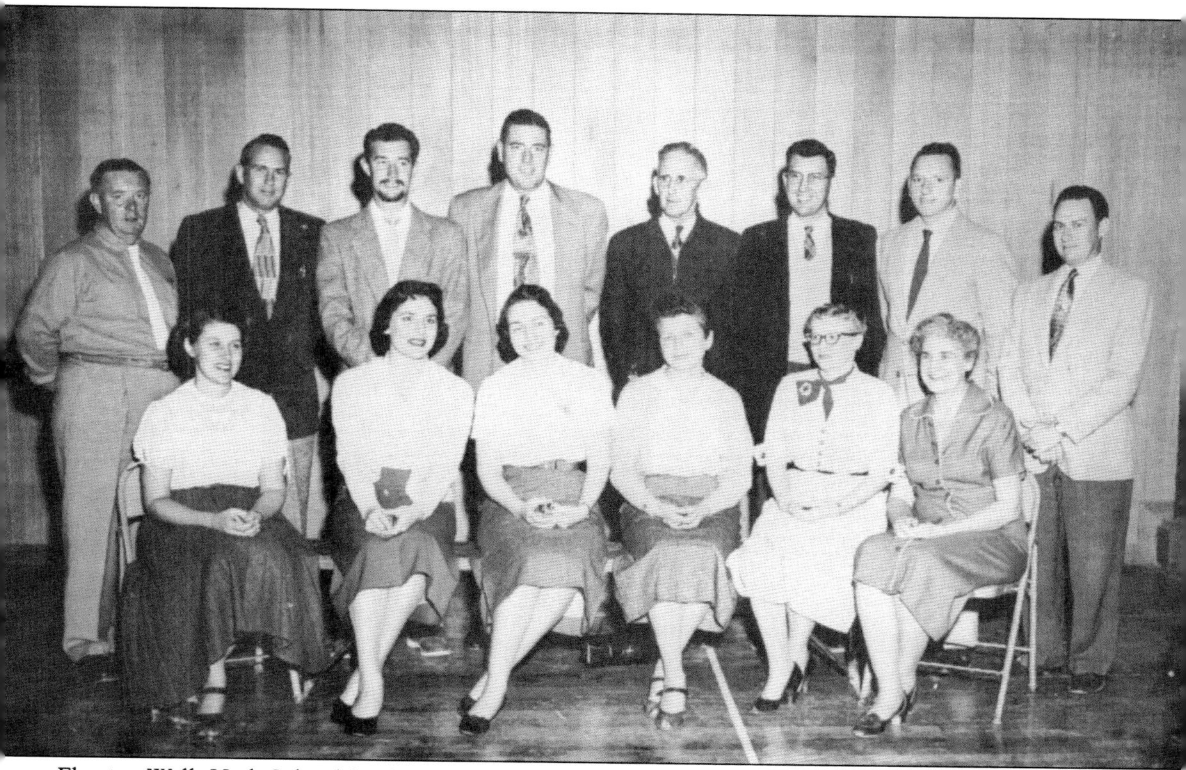

Flowing Wells High School faculty pictured here are, from left to right, (first row) Margaret Ingram, Alice Tremblay, Carol Ellsworth, Ruth Farmer, Noreen Ingle, and Joyce Thomas; (second row) Wilber Turner, A.C. Williams, Duane Miller, Eugene Moffett, Herbert Barber, Julian Aguilu, Superintendent George Smith, and Harry Spaulding. Miller was hired in 1951 as the band teacher. The first four band members were Doug Howell, Ron Howell, Carol Freer, and Marvin Fewell. Miller said, "Every time a kid stuck his head in the door, I stuck a brass drum-beater in his hand." By the beginning of school, Miller had recruited 18 band members with that method, according to W.L. Woods Jr. Williams was hired in 1950 to build up the athletic program at Flowing Wells. Rudy Reuser (FWHS 1957) remembers that Williams led the efforts to erect the original wooden light poles for the football field. Rudy was in Aguilu's Spanish class in 1953–1954, when students picked the mascot. They did not want to be an animal and liked the ideas for something "honorable and classy." "Caballero" was chosen. Joan Udell (FWHS 1957) remembers listening to Linda Sutler (FWHS 1957) write the words and music of the fight song. Irene Tompkins is credited with recommending the school colors, gold and blue. (FWUSD.)

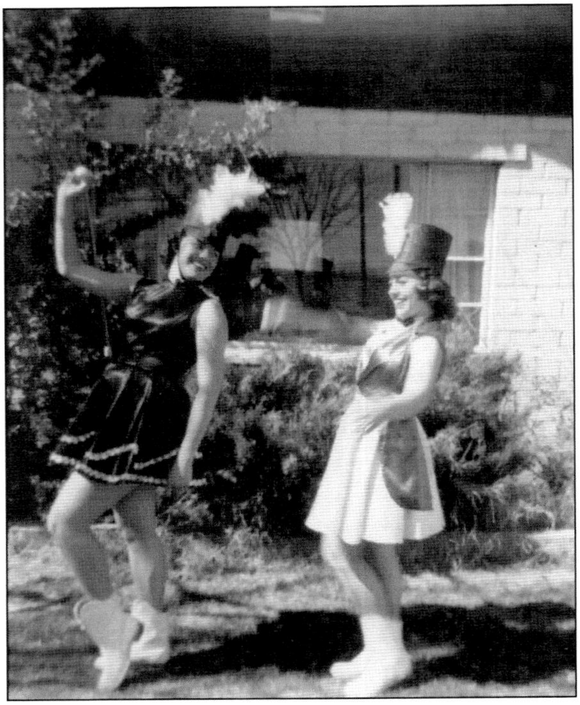

Twins Jan (left) and Jean Gardner (FWHS 1959) pose in their majorette uniforms in front of their home on Gardner Lane and Romero Road. On Wednesday, November 26, 1969, at about 9:00 p.m., residents had some excitement when a Navy A-4 jet crashed at Gardner Lane and Miller Avenue, right behind the Gardner home. Marcia remembers all the commotion resulting from the crash. The jet experienced hydraulic failure on approach for landing at Davis-Monthan Air Force Base. The pilot, Lt. William Bertsh, was able to point at an open field, eject at a mere 400 feet, and parachute to safety, with his plane crashing in a ball of flame about 100 yards from homes. His ejection seat landed 150 feet from the crash site, with the plane burrowed several feet into the ground. (Marcia Gardner Claridge.)

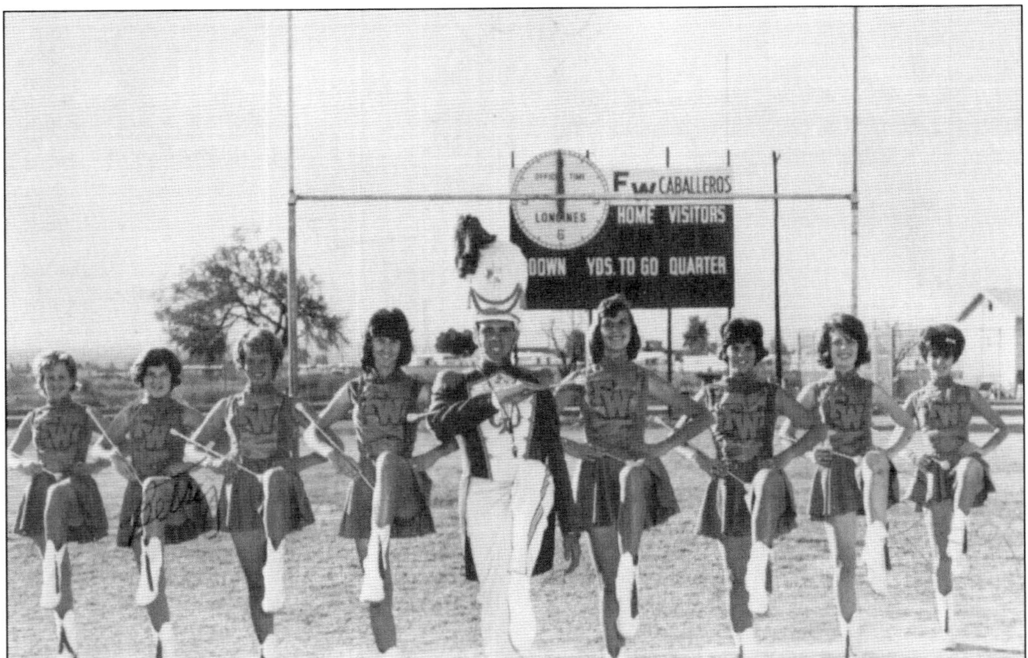

Baton twirlers perform on the FWHS football field in front of the first scoreboard in 1964. From left to right are Marilyn Gardner, Betsy Beavers, Kay Parker, Sharon Shaw, Lawrence Donny, Pam Denman, Jan Meachen, Margo Covey, and Peggy Martin. (FWUSD.)

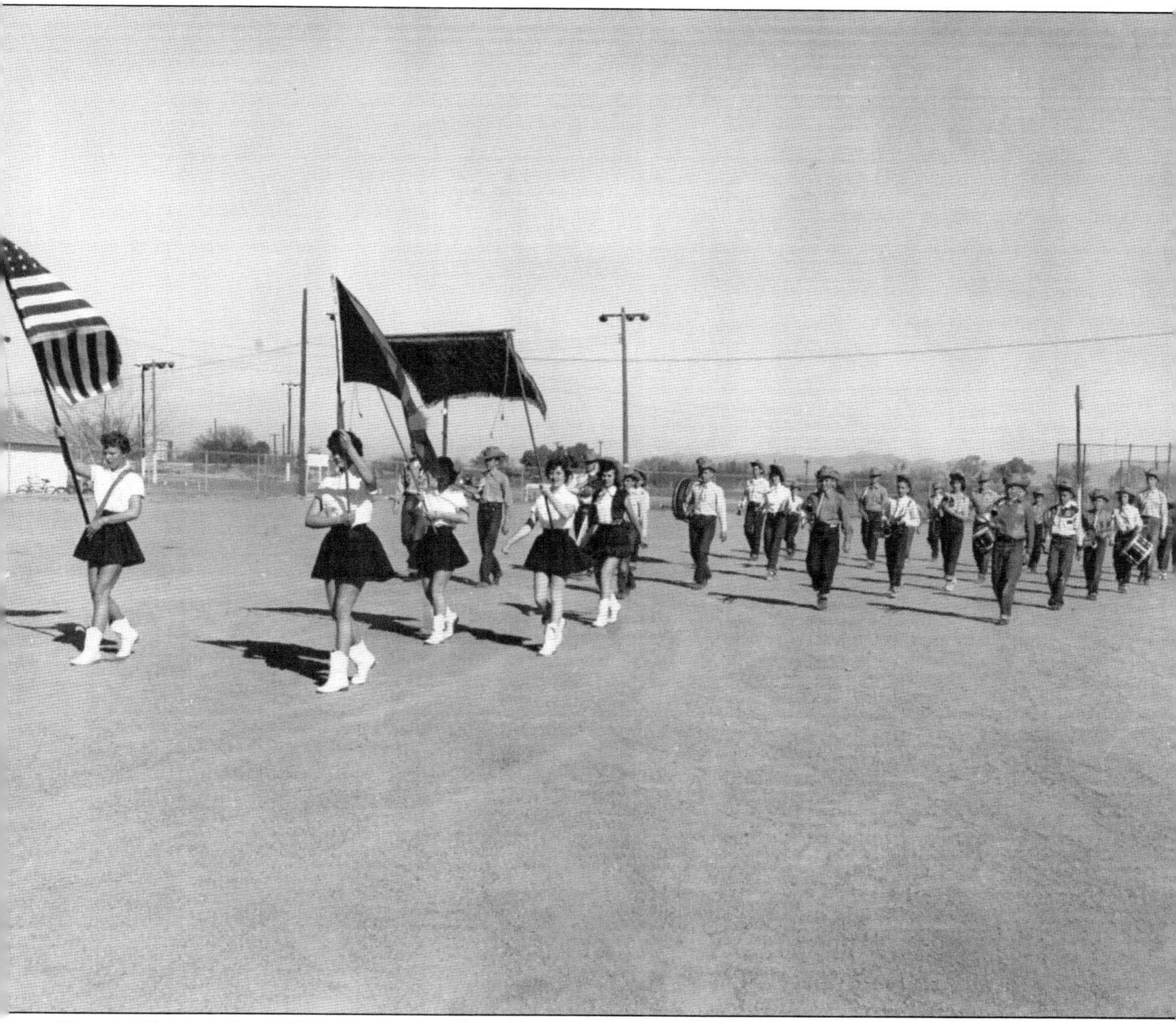

Pictured here is the 1953–1954 FWHS band marching in the February 1954 Tucson Rodeo Parade. Jimmy Elliott (FWHS 1958) identified two of the trombone players in the first row as Wilson Gee (FWHS 1957) and Dudley Westerfield (FWHS 1958). His brother Jon Elliott (FWHS 1960) is the saxophonist second from left in the second row. The saxophonist at far left in the third row is Cloise Barker (FWHS 1960). The first band was formed in 1951 under director Duane Miller. The members were Doug Howell and Norris Gomph, trombone; Carol Freer, Sunny Irwin, Jimmy Elliott, Martha Dobias, Bonnie Barthold, Jim Staples, and Ronald Loomis, clarinet; Bill Busacker, Ronald Howell, Dave Dobias, and John McKinney, trumpet; Arlie Gomph and Marvin Fewell, bass drum; Buster Steinback and Dale Howe, tenor sax; Art Smith, melophone; and Dottie Geeting, drums. (FWUSD.)

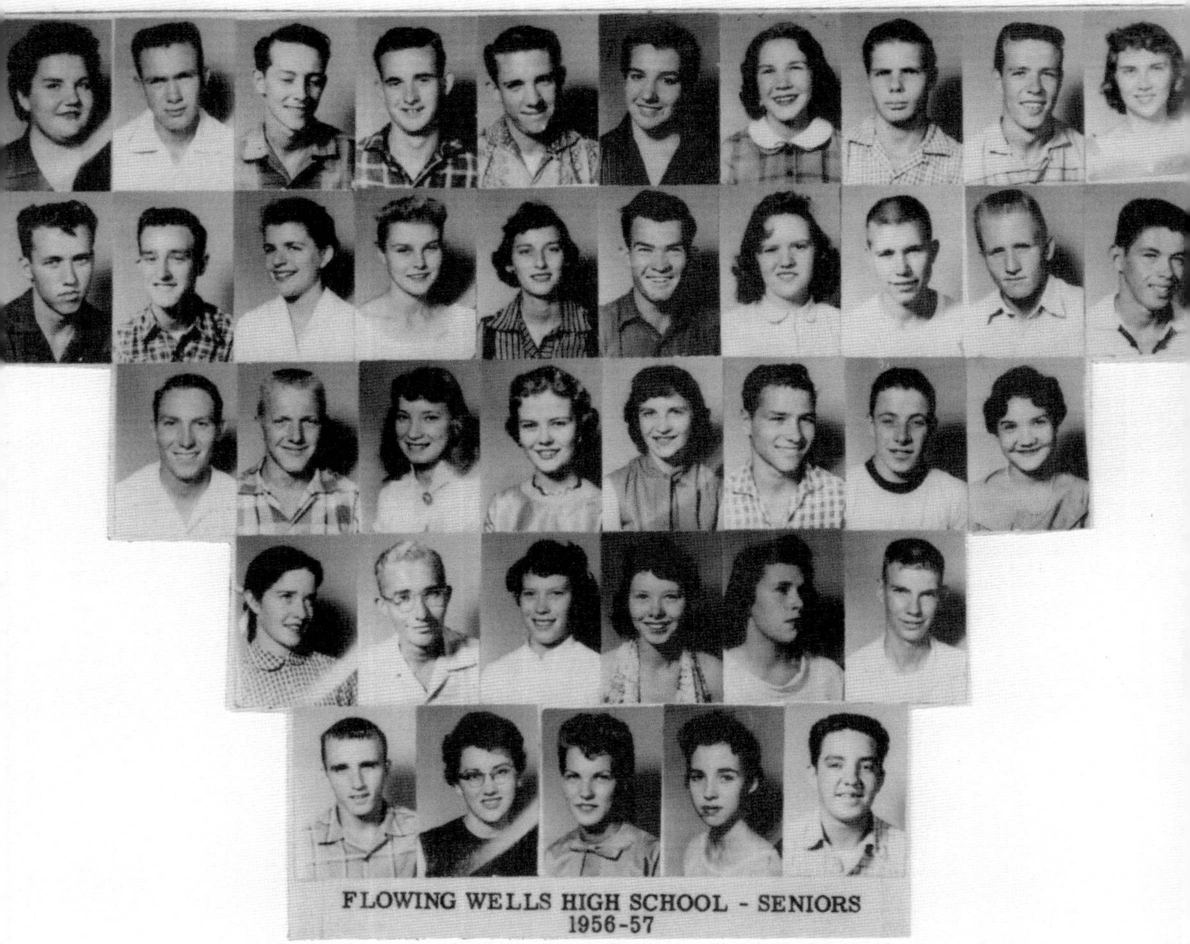

FWHS's first 12th-grade class graduated on May 28, 1957. From left to right are (top row) Pat Monroe, Robert Palmer, Milton Burnett, Pat McGuire, Wilson Gee, Bonnie Barthold, Geraldine Thomas, Frank DeSchutter, Kenny Deitering, and Joan Udell; (second row) Bob Roark (who moved before graduation), Ray Banks, Elizabeth Chase, Sherry Gates, Shirley Sipe, Don Stevenson, Judith Pulver, Art Smith, Harland King, and Bill Burgess; (third row) Frank Russell, Bill Busacker, Louise Younger, Linda Sutler, Birdie Geeting, Stuart Criss, Mitchell Chadwick, and Yolanda Contreras; (fourth row) Susie McCook, Terry Tombaugh, Marilyn Ketchem, Barbara Newman, Pat Monett, and Harry Schneider; (bottom row) Ronnie Ammerman (not in the graduation program), Jean Woods, Mary Smith (who moved before graduation), Iva Jean Lightner, and Rudy Reuser. Graduates not pictured are Leah K. Turner, Mark Peters, and Carole Ritter. (FWUSD.)

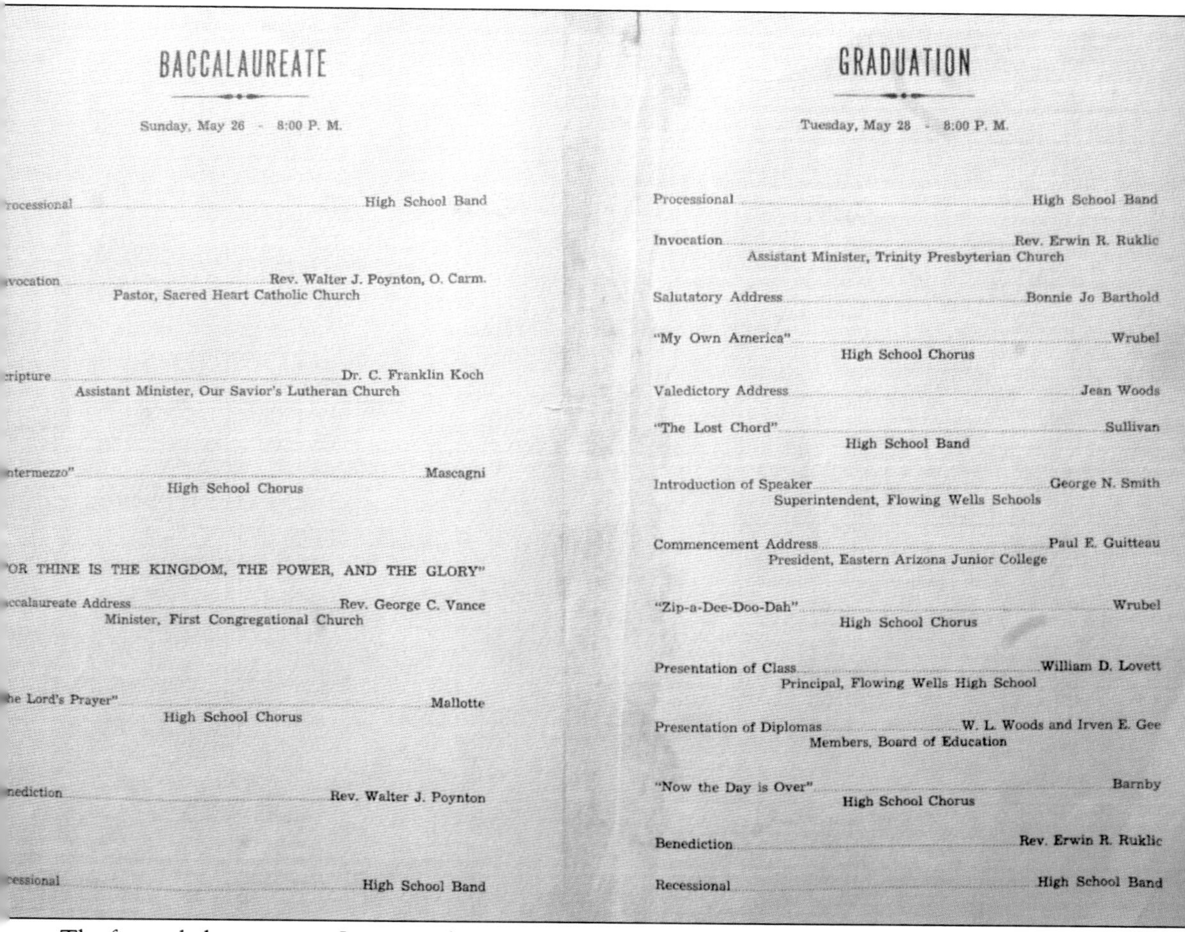

BACCALAUREATE

Sunday, May 26 - 8:00 P. M.

rocessional .... High School Band

ivocation .... Rev. Walter J. Poynton, O. Carm.
Pastor, Sacred Heart Catholic Church

cripture .... Dr. C. Franklin Koch
Assistant Minister, Our Savior's Lutheran Church

ntermezzo" .... Mascagni
High School Chorus

'OR THINE IS THE KINGDOM, THE POWER, AND THE GLORY"

ccalaureate Address .... Rev. George C. Vance
Minister, First Congregational Church

he Lord's Prayer" .... Mallotte
High School Chorus

nediction .... Rev. Walter J. Poynton

cessional .... High School Band

GRADUATION

Tuesday, May 28 - 8:00 P. M.

Processional .... High School Band

Invocation .... Rev. Erwin R. Ruklic
Assistant Minister, Trinity Presbyterian Church

Salutatory Address .... Bonnie Jo Barthold

"My Own America" .... Wrubel
High School Chorus

Valedictory Address .... Jean Woods

"The Lost Chord" .... Sullivan
High School Band

Introduction of Speaker .... George N. Smith
Superintendent, Flowing Wells Schools

Commencement Address .... Paul E. Guitteau
President, Eastern Arizona Junior College

"Zip-a-Dee-Doo-Dah" .... Wrubel
High School Chorus

Presentation of Class .... William D. Lovett
Principal, Flowing Wells High School

Presentation of Diplomas .... W. L. Woods and Irven E. Gee
Members, Board of Education

"Now the Day is Over" .... Barnby
High School Chorus

Benediction .... Rev. Erwin R. Ruklic

Recessional .... High School Band

The first valedictorian was Jean Woods, and the first salutatorian was Bonnie Barthold. Graduation programs today still follow the same basic format as this 1957 program. The baccalaureate today is more of a celebration, with students and the principal making speeches and presenting a slide show of the graduates. Today's ceremony not only celebrates graduates but also inducts outstanding alumni into the hall of fame with speeches from those alumni. Construction started on FWHS in June 1955, and students moved into the school in March 1956. Five seniors graduated in 1956: Barbara Burzyck, Charles Clausen, Duard Grove, Jane Horner, and Jan Yarian. (FWUSD.)

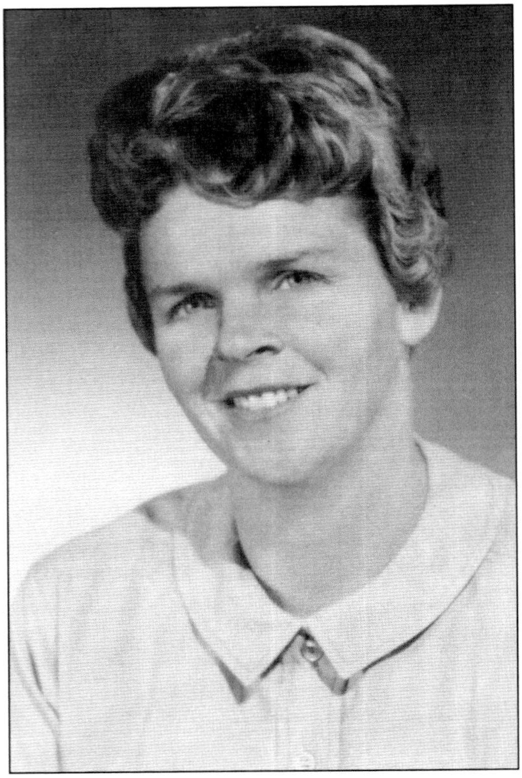

Mary Holloway joined Flowing Wells in 1956 as the girls' coach for seven sports, also serving on the Girls Athletic Association from 1956 to 1972. After Title IX in 1972, she served as the girls' coach for volleyball, softball, and gymnastics until 1982. Flowing Wells won state gymnastics championships from 1967 to 1969 and the AA Division championship in 1976–1977 and won state and division track and field championships in 1976–1977. Holloway earned a bachelor of science degree in education (1951), a master's in education (1956), a school counselor certificate (1974), and an administrative certificate (1981), all from the University of Arizona. From 1974 to 1990, she was a member of the guidance and counseling staff at Flowing Wells, and was the first women's athletic director in the state from 1981 to 1992. She was elected to the Pima County Sports Hall of Fame in 1992 and Flowing Wells High School Athletic Hall of Fame in 2003. Since 1990, she has served as the main driver and consultant for the Flowing Wells Alumni Association. (FWUSD.)

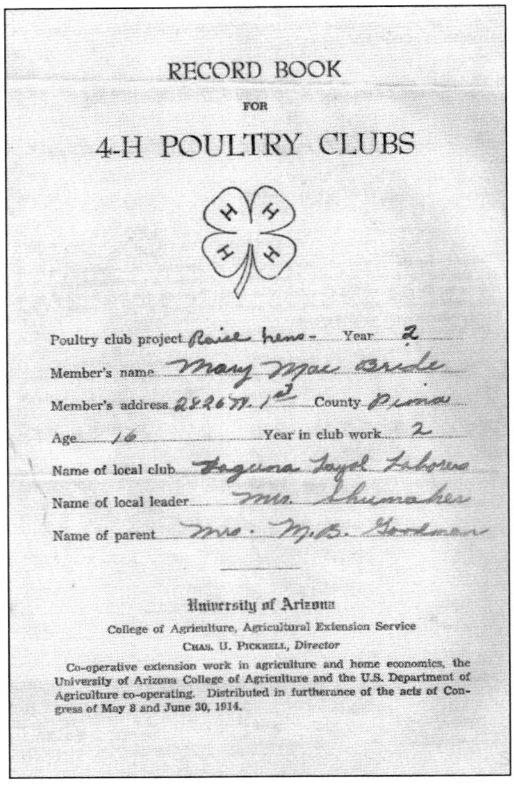

RECORD BOOK

FOR

4-H POULTRY CLUBS

Poultry club project Raise hens - Year 2

Member's name Mary Mac Bride

Member's address 2826 N. 1st County Pima

Age 16 Year in club work 2

Name of local club Laguna Loyal Laborers

Name of local leader Mrs. Shumaker

Name of parent Mrs. M. B. Goodman

University of Arizona

College of Agriculture, Agricultural Extension Service

Chas. U. Pickrell, *Director*

Co-operative extension work in agriculture and home economics, the University of Arizona College of Agriculture and the U.S. Department of Agriculture co-operating. Distributed in furtherance of the acts of Congress of May 8 and June 30, 1914.

Before Mary Holloway taught at Flowing Wells, she was a Laguna Loyal Laborer as Mary MacBride at 16. The Loyal Laborers were a club at Laguna School created by Susan Edna Shumaker to train children in citizenship and individual development as part of 4-H. (FWUSD.)

FWHS is seen from the air in 1957. At top are the north fields, where the agriculture, baseball, practice, and girls' softball fields are currently. This land was previously owned by the Lawson family, who grew cotton and whose home was at the corner of Flowing Wells Road and Roger Road, where Food City sits today. The football field was equipped with bleachers, lights, and a scoreboard. Flowing Wells Road was a two-lane dirt road at the time. (FWUSD.)

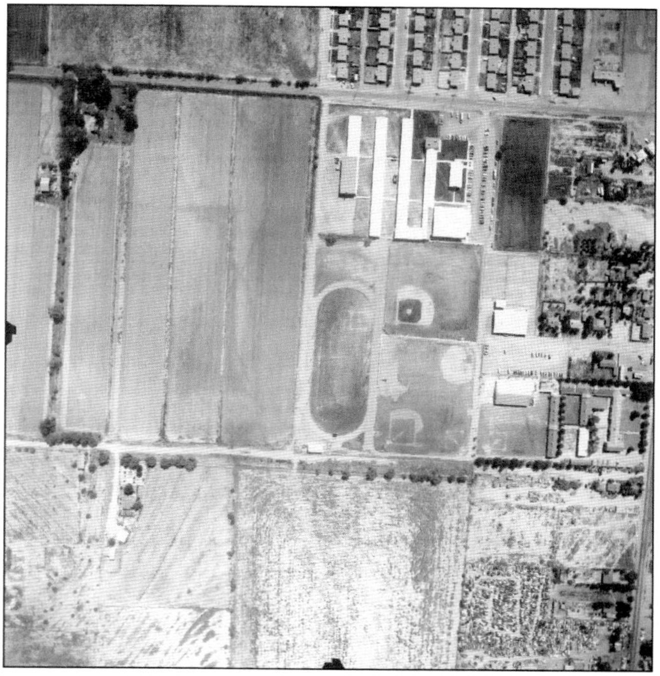

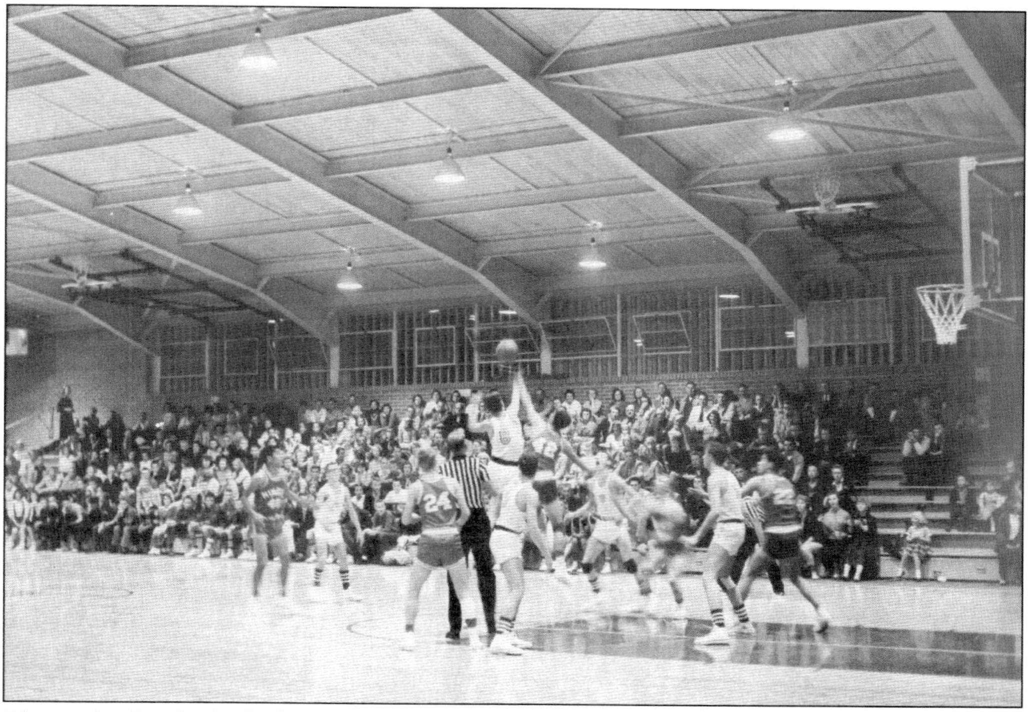

This is the first basketball game played in the new gym at FWHS on December 17, 1957. The Caballeros, coached by Lionel Goar (driver training, physical education, and social studies), won 55-35 against San Manuel High School. Today, this is known as the east gym and is home to the Lady Cabs basketball team. The west gym, built in 1983, hosts girls' and boys' sports as well as assemblies and other activities. Goar was inducted into the Flowing Wells High School Athletic Hall of Fame in 2009. (FWUSD.)

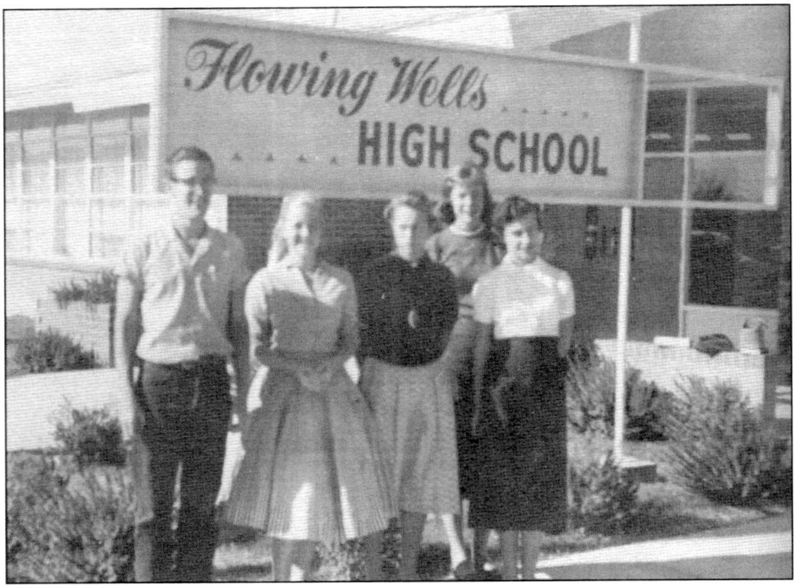

The 1959 Future Teachers of America Club is pictured in front of the FWHS sign. From left to right are Paul Dayton, Firelei Douglas, Bonnie Cornett, Laura Newsome, and Zelma Wilson. (FWUSD.)

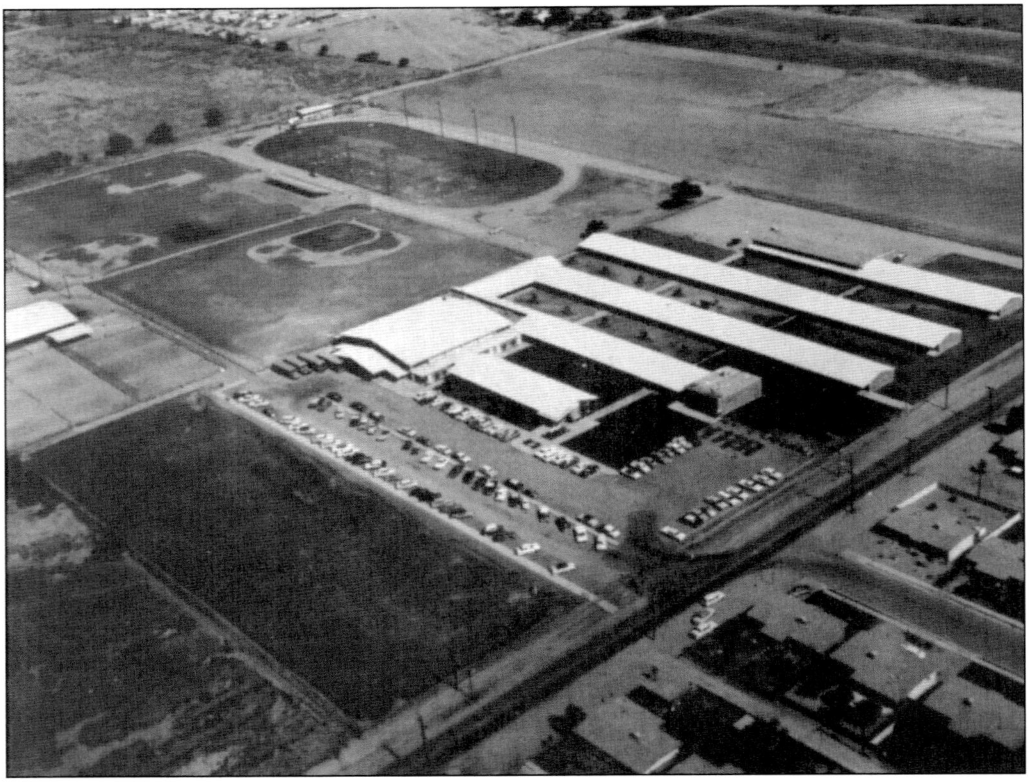

This c. 1963 aerial view shows FWHS. The auditorium was built just south of the parking lot on Flowing Wells Road in 1967, and the library was built in 1968. The current cafeteria, bookstore, and additional classroom buildings to the west of the auditorium were built around 1998. The words of the school alma mater are: "All hail to Flowing Wells, all hail gold and blue. We never will forget you all our whole life through. All hail to Flowing Wells, all hail to friendship true. Our loyalty we give to you, all hail to Flowing Wells." (FWUSD.)

Vic Meneley (1920–2012) moved to Tucson in 1961, becoming assistant principal of FWHS. He was principal from 1967 until 1975, when he became director of secondary education. He retired in 1981. He loved sports and was inducted into the Flowing Wells High School Athletic Hall of Fame in 2007. After his retirement, he played basketball on Sundays in the high school gym. He played sports into his 80s. Meneley was talented at performing arts, appearing in a variety of shows, and was a baritone in the Tucson Barbershop Chorus. (FWUSD.)

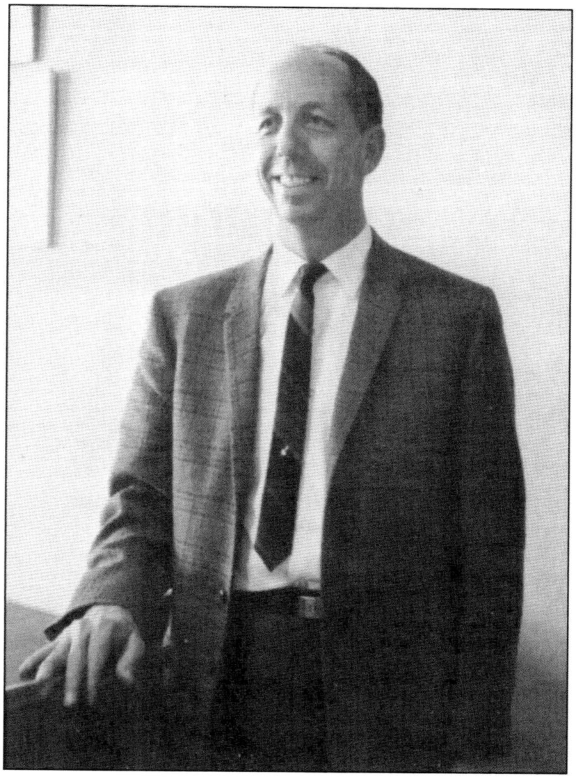

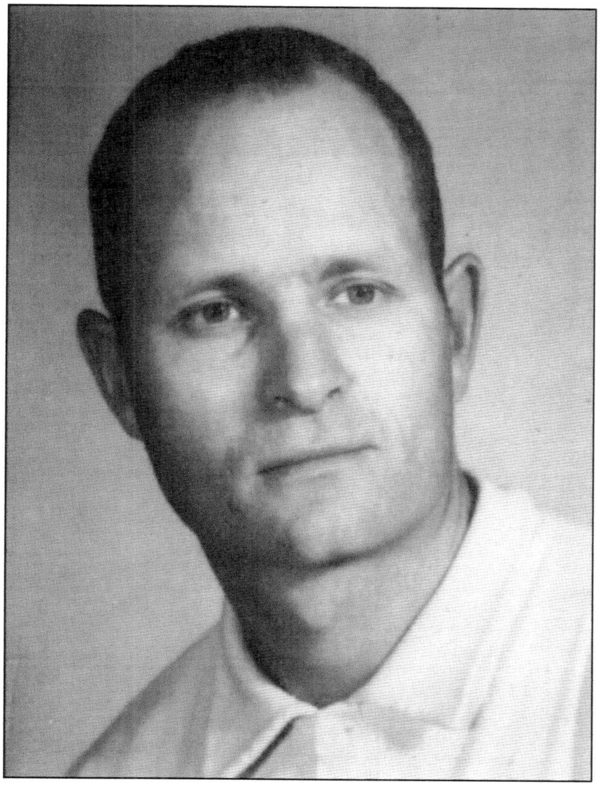

Larry Hart coached football at FWHS from 1963 to 1965 and from 1968 to 1979. He took over after the city's first ever 0-10 season in 1962. In 1964 and again in 1968, he led the Caballeros to the state championship. Hart was named assistant principal in 1975. That year, the Caballeros won the AA state championship against McClintock High, 14-13, clinching a perfect 10-0 record. Hart had six state championship trips and 146 total wins. He was inducted into the Arizona Coaches Association Hall of Fame in 1984, and in 2003, he was part of the first class inducted into the 2003 Flowing Wells High School Athletic Hall of Fame. Ted Sorich was another significant figure in Flowing Wells starting in 1966. Sorich was inducted into the FWHS Athletic Hall of Fame in 2005. (FWUSD.)

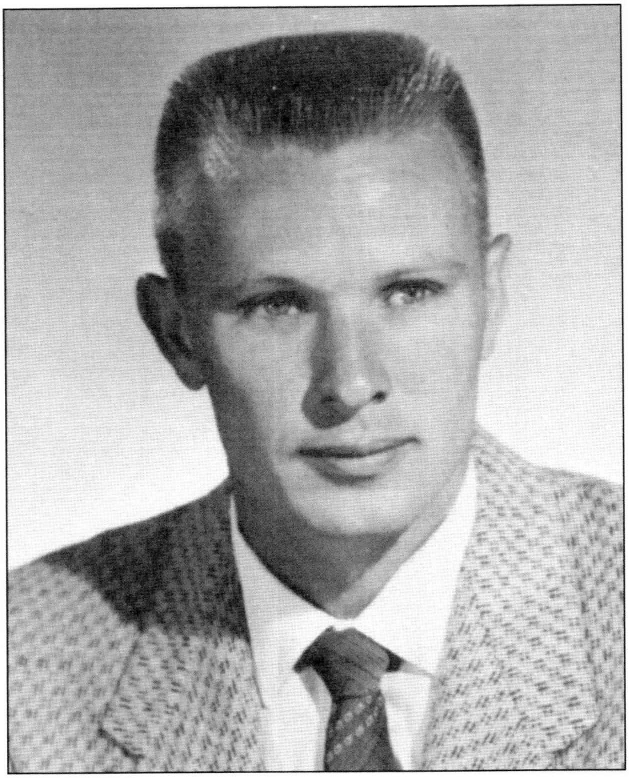

Richard "Dick" Griesser (1932–2009) started at FWHS in 1959 as a physical education teacher. He served as assistant principal from 1981 until his retirement in 1988. Griesser was an All-American outfielder for the University of Arizona (playing for Frank Sancet) in three College World Series. He played for the US Olympic baseball team in 1956 in Melbourne, Australia. He was part of the Tucson sports scene for more than 40 years. He was inducted into the University of Arizona Sports Hall of Fame and Flowing Wells High School Athletic Hall of Fame in 2007. (FWUSD.)

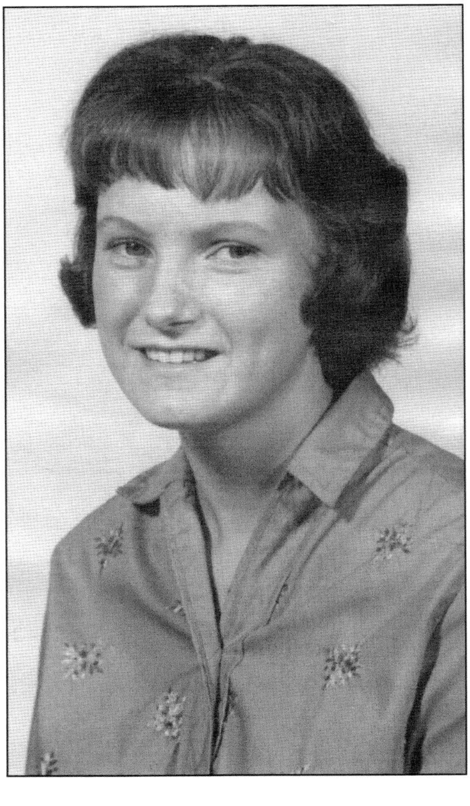

Jeretta Rice Douglas (FWHS 1961) married Walter Douglas III (FWHS 1960) in 1967. Jeretta still lives in the house they built on Roger Road. A cosmetologist by trade, she operates her salon, Jeretta's Hair Care Center. In 2002, she was given the All Arizona School Board Member Lifetime Achievement Award. In 2008, she received the Barbara Robey Lifetime Achievement Award from the Arizona School Boards Association. Jeretta served on the school board from 1986 to 2016. Her father-in-law, Walter Douglas Jr., served on the same board from 1942 to 1987, meaning a Douglas served on the Flowing Wells School Board for 75 consecutive years. (Jeretta Douglas.)

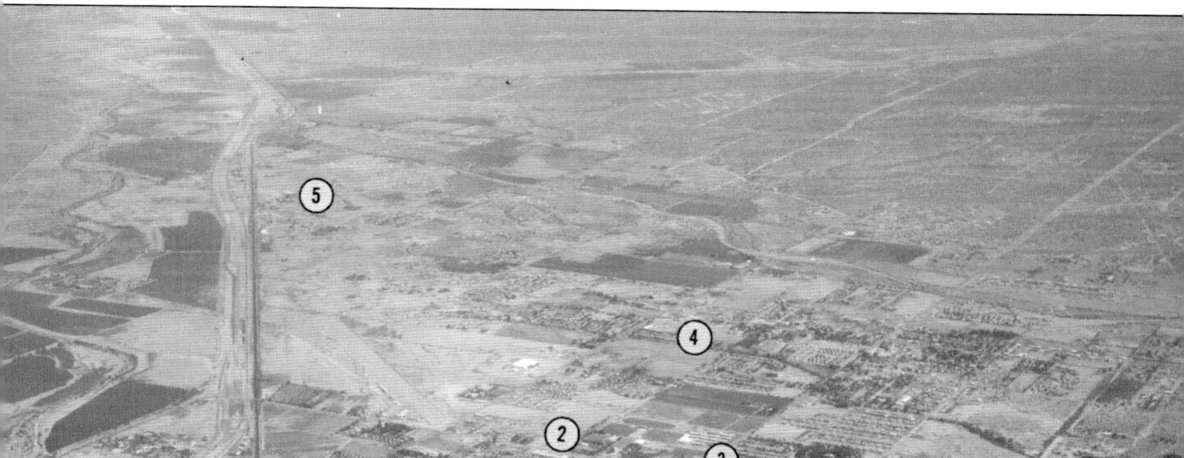

This bird's-eye view shows the Flowing Wells area in 1964. There were five schools in the district at the time: Walter Douglas Elementary (1963), Iola Frans Elementary–Flowing Wells Junior High (originally Flowing Wells Elementary School, 1940), Flowing Wells High School (1956), Homer Davis Elementary (originally Romero Road School, 1955), and Laguna Elementary (1889). Additional schools were built as the area grew in population: Inscape Alternative (1975) on Gardner Lane was renamed Sentinel Peak in 2003. In 2010, Sentinel Peak High School moved to its new location on Aerie Drive. A new Flowing Wells Junior High–McClelland Educational Park was built at Wetmore Road and La Cholla Boulevard in 1976 on land partially donated by the McClelland family and previously used as feed lots for Shamrock Dairy. Robert Richardson Elementary (1980) was named for a longtime Flowing Wells board member. Centennial Elementary was built in 1990, followed by J. Robert Hendricks Elementary (1997), named for a longtime principal and superintendent, and the Emily Meschter Early Learning Center (2010), named to honor a devoted friend of the FWUSD staff and students. (FWUSD.)

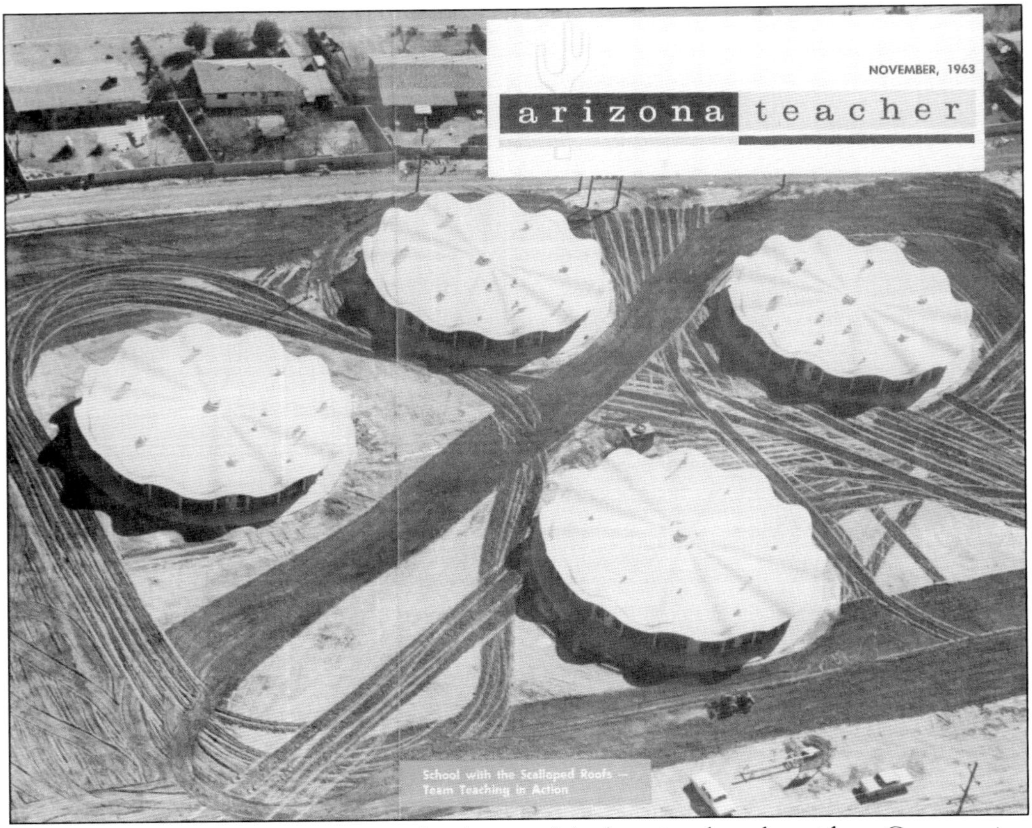

Walter Douglas Elementary was named in honor of the longtime board president. Construction started in 1961, and John Shaver and Co. of Salina, Kansas, designed the circular buildings where all the classrooms opened up to the center. Reid and Hazard of Tucson served as the associate architects. The open-classroom, team-teaching concept was revolutionary and award-winning and is still used today in education. By the late 1960s, movable walls were added between classrooms, and more traditional walled classrooms proved to be the preferred configuration. (FWUSD.)

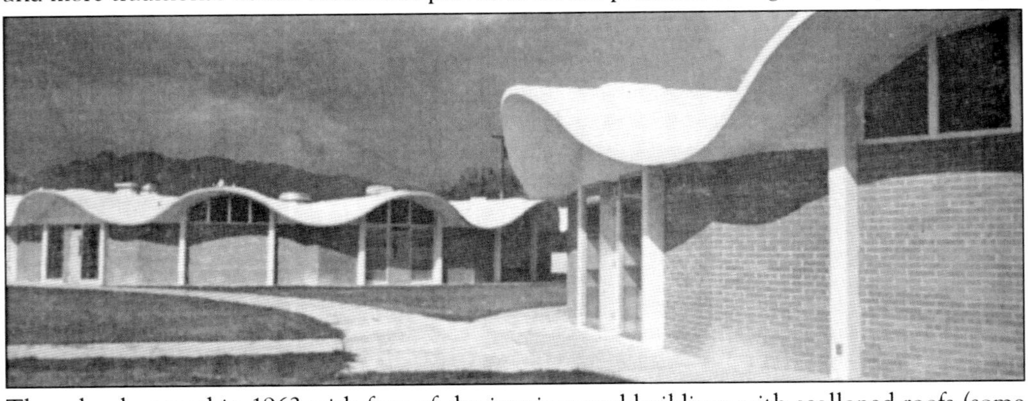

The school opened in 1963 with four of the iconic round buildings with scalloped roofs (some locals refer to them as "wavy" or "potato chip") and a new, highly publicized, open-design team-teaching philosophy. Superintendent George N. Smith, a doctoral candidate at Stanford University in 1960, concentrated his studies on school buildings and construction. With a $15,000 grant from the Ford Foundation, the concept of Walter Douglas Elementary became a reality. Two additional buildings of the same design were added in 1967, bringing the total to six. (FWUSD.)

Walter Douglas Elementary School teacher Mona Dayton was named the 1966 National Teacher of the Year. Pres. Lyndon B. Johnson presented the award to her on April 5, 1966, at the White House. (FWUSD.)

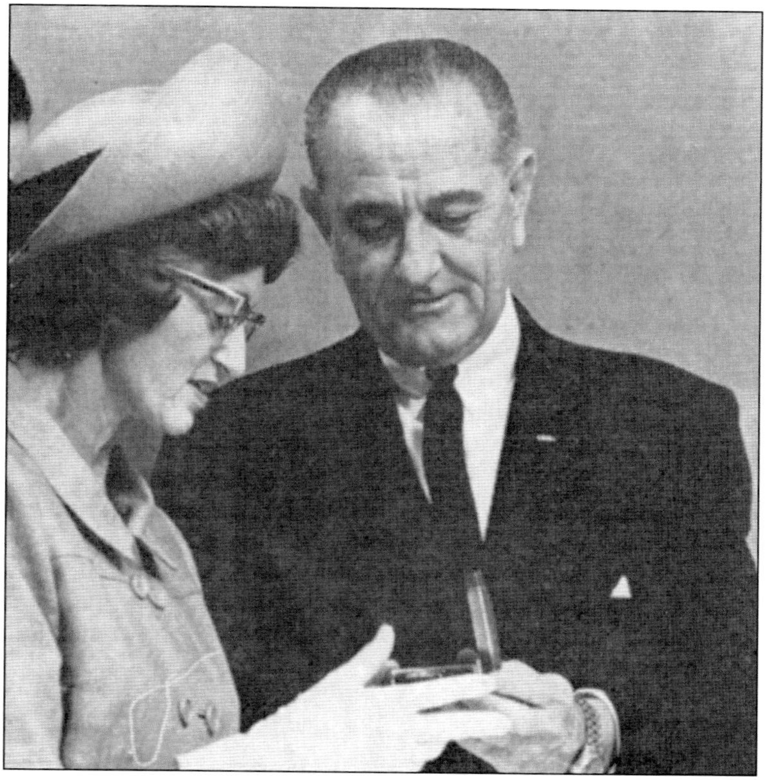

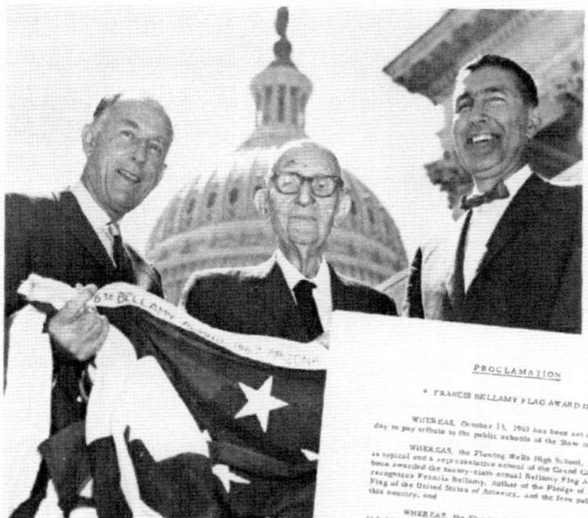

Senators Paul Fannin and Carl Hayden with Congressman Morris Udall proudly display the 1967 Bellamy Award flag, which was flown over the Capitol May 18, 1967.

PROCLAMATION

" FRANCIS BELLAMY FLAG AWARD DAY "

WHEREAS, October 13, 1967 has been set aside as a special day to pay tribute to the public schools of the State of Arizona, and

WHEREAS, the Flowing Wells High School, Tucson, which is typical and a representative school of the Grand Canyon State, has been awarded the twenty-sixth annual Bellamy Flag Award which recognizes Francis Bellamy, author of the Pledge of Allegiance to the Flag of the United States of America, and the free public schools in this country, and

WHEREAS, the Flowing Wells High School has been designated to hold the honor for a fifty-year period in the group of outstanding and representative Bellamy Award schools throughout the nation, and

WHEREAS, the Bellamy Advisory Board has judged the representative Flowing Wells High School as one with these characteristics contributing to its choice: "the proficient performance of duty by the administration, a loyal and well-balanced staff, the emphasis on school spirit, patriotism, and citizenship, the philosophy: 'The Pursuit of Excellence,' a responsible student body, a solidified community spirit, the curriculum, recognition of the worth of field trips and visits to educational areas, the student council projects, and the accomplishments of the alumni."

NOW, THEREFORE, I, Jack Williams, Governor of the State of Arizona, do hereby proclaim Friday, October 13, 1967 as

FRANCIS BELLAMY FLAG AWARD DAY

in the State of Arizona.

IN WITNESS WHEREOF, I have hereunto set my hand and caused to be affixed the Great Seal of the State of Arizona.

DONE at the Capitol in Phoenix this 11th day of May in the year of Our Lord One Thousand Nine Hundred and Sixty Seven and of the Independence of the United States the One Hundred and Ninety-first.

GOVERNOR

ATTEST:

Secretary of State

FWHS received the 26th Bellamy Award in 1967. This award was one of the highest offered to institutions of secondary education. The ceremony was held at the high school on October 13, 1967. Two Western Union telegrams were sent to William Lovitt, assistant superintendent of Flowing Wells Public Schools, on May 16 and 17, 1967, from Vice Pres. Hubert H. Humphrey and Pres. Lyndon B. Johnson. Ronald and Nancy Reagan spoke at the high school on Bellamy Day, October 26, 1979. On April 23, 1968, Sen. Barry M. Goldwater, the 1964 Republican presidential nominee, spoke in the high school auditorium with a sign over the stage reading "A peacemaker school welcomes a peacemaker." The school district won the National Education Association's 1964 Peacemaker Award. (FWUSD.)

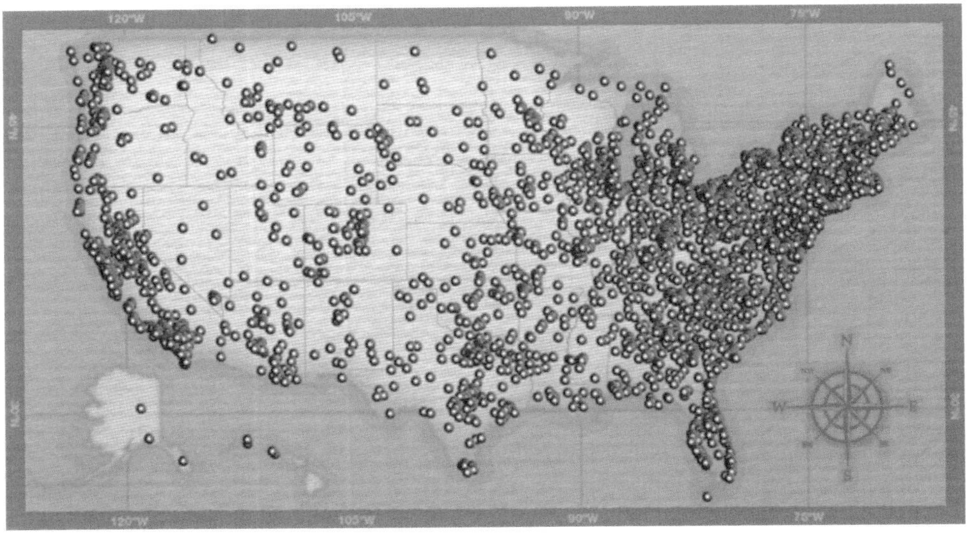